Charles Hose

A Descriptive Account of the Mammals of Borneo

Charles Hose

A Descriptive Account of the Mammals of Borneo

ISBN/EAN: 9783337245689

Printed in Europe, USA, Canada, Australia, Japan

Cover: Foto ©ninafisch / pixelio.de

More available books at **www.hansebooks.com**

A

DESCRIPTIVE ACCOUNT

OF THE

MAMMALS OF BORNEO.

BY

CHARLES HOSE, F.R.G.S, F.Z.S,

Fellow of the Anthropological Institute;

Member of the British Ornithologists Union,

RESIDENT OF THE BARAM DISTRICT, SARAWAK.

LONDON.

1893.

INTRODUCTION.

The following DESCRIPTIVE ACCOUNT OF THE MAMMALS OF BORNEO, *has been based mainly on the collections made by myself in the Baram District of Sarawak during the years from 1884 to 1892. The map appended to this Account indicates the localities at which these collections have been made by having their names underlined. In preparing these pages I have to acknowledge the valuable assistance I have received from the following authoritative works on the Mammals occurring in this part of the Malayan region: from Dr. John Anderson's* ZOOLOGY OF WESTERN YUNNAN; *from Papers by Dr. Jentink in the well known* NOTES FROM THE LEYDEN MUSEUM; *from Dr. Blanford's* FAUNA OF BRITISH INDIA, *Dr. T. Horsfield's* RESEARCHES IN JAVA, *and Dr. Dobson's* CHIROPTERA. *I have also taken full advantage of the various writings of Drs. Müller and Schlegel. But to no one more than to Mr. Oldfield Thomas of the British Museum, am I indebted not only for the aid I have derived from his numerous papers in the* PROCEEDINGS OF THE ZOOLOGICAL SOCIETY *and in the* ANNALS AND MAGAZINE OF NATURAL HISTORY *describing my own and other Bornean collections; but more than I can sufficiently acknowledge, to his personal assistance and advice so generously and heartily afforded at all times. It was on his suggestion indeed that I have been prevailed on to attempt the task of putting together for private circulation in collected form the information on the Mammalia of Borneo scattered through the various publications I have just named, to which I have added such personal observations on their habits as I have had the opportunity of making in the field. My object in undertaking this task—of whose many shortcomings I am very sensible—has been to supply a handy guide to the Mammals of the country in which I have spent so many years, in the hope that it may be of some assistance to those engaged in the study of the fauna of this large and interesting island.*

I cannot close these remarks without expressing how much I am indebted to my good friend Mr. A. H. Everett, not only for his having induced me to take up the investigation of the fauna of Borneo, but for the invariably kind manner in which he has assisted me throughout my researches.

CHARLES HOSE.

PREFACE.

This work on the Mammals of Borneo has been prepared in the course of the last few years by Mr. Charles Hose, an enthusiastic naturalist, who, in conjunction with his friend, Mr. A. H. Everett, has revolutionized our knowledge of the vertebrates of the Island of Borneo, or at least of that part of it ruled by His Highness The Rajah of Sarawak, in whose service he is, and by whose active encouragement he has been enabled to carry on his researches into the fauna. These researches, necessarily carried out on a large scale, have already resulted in such an increase of our knowledge that it is hardly too much to say that the mammals of the wild and previously unexplored Baram district of Sarawak, are now as well-known as are those of any other district in the world of equal size and richness, and this is entirely due to the energy and enthusiasm of Mr. Hose.

The descriptions in this work are of course largely compiled from the works of various technical writers, but the notes on the habits, native names, and distribution are all Mr. Hose's own, and are based both on his really intimate acquaintance with the animals themselves, and also on his extensive knowledge of the natives, to whom, as is the case with all successful collectors, he is largely indebted for information and specimens.

The determinations throughout the book are practically all my own, and as I am therefore responsible to a certain extent for its correctness, Mr. Hose has asked me to write these few words by way of preface, but it is difficult to give persons unacquainted with him any idea of his energy, scientific spirit, and intimate knowledge of his subject, of which the outcome is the present useful little work. This we have every reason to hope is only the precursor of a fuller and more complete account, which he hopes to prepare if he is able in the future as in the past to carry on his painstaking researches into the Natural History of the mammalian inhabitants of the Island of Borneo.

OLDFIELD THOMAS.

THE MAMMALS OF BORNEO.

ORDER PRIMATES.

FAMILY SIMIIDÆ.

GENUS SIMIA.

SIMIA SATYRUS.

Simia Satyrus, Linn., Syst., Nat. 1766, p. 34.

The bare cheeks are enormously outwardly enlarged, the skin of the face and of the great bare area on the guttural sacks being livid black. The skin of the face is sparsely covered with short red hairs, and the forehead also is almost naked. The sides of the upper lip and the chin are clad with long bright maroon-red hairs. The hair on the middle of the head, immediately behind the forehead, is rather short, while that over the temporal and parietal regions is long and directed forwards. On the rest of the head the hair is dark maroon-brown, this colour also extending round the sides of the neck and on to the throat. Between the shoulders the colour is more rufous, whilst down the back it is almost as dark as the head, the sides being also maroon red, as well as the shoulders, the arms being almost red, and thus paler than every other part of the body. The lower portions of the thorax and the abdomen are dark chestnut-red. The legs are almost as pale as the arms. The hair on the body generally, and on the limbs is very long, measuring as much as 13 to 14 inches.

The skull has well-developed sagittal and lambdoidal ridges, and the orbital ridges are also well marked, and the malo-maxillary area is broad. Dyak name, 'Maias.' Kayan, 'Balli Poi.'

Hab. Sadong River (C. S. Pearse). Sarawak (G. D. Haviland). Simunjan (J. Revilliod).

GENUS HYLOBATES.

HYLOBATES MÜLLERI.

Hylobates mülleri, Martin. Nat. Hist. Quadr. 1841, p 444;

This species varies from grey to dark yellowish brown, but the grey in certain lights appears pure ashy, and in others of a brownish tint. In some the chest and abdomen are frequently of a lighter colour than the other parts, and of a brownish yellow, and this seems to be the character of individuals met with on the west coast of Borneo, while those inhabiting the meridional parts of the island have the hands and fore part of the body of a black-brown or reddish-brown. In both of these varieties there is a yellowish white supercilium. The last of them leads into the Hylobates from the neighbouring islands of Sulu to the north-east of Borneo, in which the upper parts of the body are either grey or brownish, the lower part of the back and the loins being a little more clear than the rest. Specimens of this gibbon procured by me at Claudetown and now in the British Museum show that the colouring in different parts of the body must be considered of little importance, as I obtained eleven specimens, five of which were in the same troop and the other six from the same locality varying in colour as much as it is possible for them to do; some had yellowish backs and black chests, others black backs with yellowish chests, and some nearly black all over, whilst others were almost a complete silver-grey. I therefore come to the conclusion that *Hylobates mülleri* and *Hylobates leuciscus* cannot be separated. The peculiar bubbling noise they make is similar. I think it very unlikely that two distinct species should be so constantly found together as they are in Sarawak.

The natives call the silver grey variety Empliau or Wa, Wa, and the dark one Empliau arang (coal) because of its colour. The noise made by these gibbons is very pretty, commencing punctually at five o'clock in the morning and continuing till the sun is above the tops of the trees. They become very tame and make very nice pets.

Hab. Baram River (C. Hose). Mount Dulit (C. Hose). Sarawak (G. D. Haviland). Batang Lupar (C. Hose).

FAMILY CERCOPITHECIDÆ.

GENUS MACACUS.

MACACUS NEMESTRINUS.

Macacus nemestrinus F. Cuv., Hist. Nat. des Mammif. Août 1820, pl. xlii. Jan. 1822, pl. xliv.

The general colour is a decided olive, tending in some animals to brown, the variation in colour being due to the relative development of the yellow and black rings on the hair. The rings occur on the exposed portion of the hair, the hidden part of which is grey. The upper surface of the head, the mesial line of the back, and the upper surface of the tail near its base, are deep brown or even blackish, more especially on the head and over the hind quarters. The extremities pale towards the hands and feet, which are light olive-brown. The outsides of the thighs have an olive-grey tint. Some animals, however, especially the fully grown ones, are almost uniformly coloured deep olive-brown except on the blackish head and the middle line of the back. The sides of the face and the under surfaces generally are greyish, tending to white; but on the sides of the face the hair is washed with a dark, almost blackish grey. The face is nude, of a dusky flesh colour, which is the tint also pervading the almost naked ears and the callosities. A few scattered hairs occur about the mouth. On the top of the head, especially on the dark coloured area, the hairs in the adult are short, rather erect and profuse; on the under parts they are rather sparse, especially on the abdomen. The muzzle is rather long and dog-like; the body is short, compact, and broad chested, with moderately long, powerful limbs. The head is somewhat flattened above, with pronounced supraorbital ridges. The limbs are relatively longer than in *M. leoninus*. The tail is little more than one-third the length of the body and head, and is rather sparsely clad, contracting somewhat to a point and carried erect, being somewhat downwardly curved near the tip.

There are not the marked variations between different specimens that are found in *M. leoninus*, Blyth, to which this form is most closely allied, the males and females being alike, and the young are only a little more richly coloured than the adults. These later attain to a great size, as is evinced by the dimensions of the cranium of the adult. I have seen specimens standing at the shoulder as high as a good sized mastiff and quite as powerful.

The chief feature of the skull is the great development of the facial portion, which is thrown much forwards.

A yellow specimen of this monkey was procured at Long Salai in the Baram district. These monkeys are easily tamed by the natives, and in some places they are used to climb the cocoanut trees to throw down the nut, the monkeys having been taught to throw down only the ripe ones. The Dyak name is 'Brok,' and the Kayan name 'Koyut.' Sometimes these animals are very savage, and will attack a dog when provoked.

Hab. Baram River (C. Hose). Mount Dulit (C. Hose).

MACACUS ARCTOIDES.

Macacus arctoides Is. Geoff. St.-Hil. Mag. de Zool. 1833, cli. pl. II. (adult).

This monkey is reported to occur in Borneo, but as yet I have not met with it.

The type of Macacus arctoides was an adult male from Cochin China, characterised by a red face, very short, stumpy tail, and by long hair, each individual hair being "*plusieurs fois annelés de brun et de roux clair.*"

A large red-faced monkey, with a stumpy tail, was purchased by the Zoological Society from a dealer in Liverpool, who could give no information regarding its habitat. After living for some years in the Society's gardens it died, and was deposited in the British Museum, where it is now stuffed, and the skeleton and skull are preserved. In its general form it is exactly like *M. brunneus*, only much larger, and it differs from it and *M. melanotus* in the general annulations of its hair all over the body, even to the under parts, which, however, are not so distinctly annulated as the upper surface.

MACACUS CYNOMOLGUS.

Macacus cynomolgus, F. Cuv. Hist. Naturelle des Mammif. Fev. 1819, Pls. 30 and 31.

The leading features of this animal are—its massive form, its large head closely set on the shoulders, its stout and rather short legs, its slender loins and heavy buttocks, and its tail thick at the base. The general colour of the monkey does not call for any remark, as it conforms to that of the other species, and it has the bluish white area internal to the eyes, and palish eyelids. This is one of the common monkeys of the Borneo jungle, found both in the low country and on the mountains to the height of 5000-ft. It is known to the natives all over Borneo as 'Kra'; this name is probably given to it on account of the noise it makes. Specimens obtained above 3000-ft. on the mountains have a tendency to become more rufous on the head than those of the low country. They are very destructive in the gardens, and the paddy crops suffer much from the depredations of these animals.

Hab. Mount Dulit 5000-ft. (C. Hose). Baram River (C. Hose). Sarawak (A. Everett).

GENUS NASALIS.

NASALIS LARVATUS.

Nasalis larvatus, Wurmb. Geoff. St.-Hil. Ann. du Mus., vol. xix., 1812, p. 90.

The upper surface of the head, neck, back and flanks dark red-brown, passing into greyish yellow on the crupper, tail, and limbs. A yellow stripe on the shoulders. The hair on the sides of the face, neck, and shoulder is long and of a yellowish tint variegated with reddish brown, and the chin is well bearded. The under parts are yellowish white, and the tail is tufted. The face is a dirty yellow merging with the white around the lips. The under surfaces of the extremities are blackish, and the ears are of the same colour, and small. Nose produced into a proboscis with large nostrils opening downwards and separated from each other by a septum, but only developed in its characteristic form at a very advanced age in both sexes, being much shorter in the young and turned upwards (*S. recurvus, Vig. and Horsf.*). The eyes are rather wide apart ; the neck short, and the throat rather swollen from the presence of a laryngeal sack.

The colours of both sexes are the same, but the female is smaller than the male. The colour as it advances in age is the subject of considerable change. In early youth the mouth and the area around the eyes are bluish, and the cheeks are curiously wrinkled, and the rest of the face is sullied brownish white, and the ear flesh coloured, but intensely black around the margin. The hands are black ; and the head with the exception of the crown, the neck, the upper part of the crest, and the front of the upper arm, are dark red-brown, the rest of the pelage being sullied, pale-yellowish brown. Through a series of changes during which the red-brown of the upper parts first increases in strength, and the grey brown of the hips and upper side of the tail change to yellowish white, the adult pelage is reached.

	ft.	in.
Length of adult, muzzle to base of tail...	2	5
Length of tail	2	2

This peculiar form, which has all the structural characters of *Semnopithecus*, appears to be restricted to the island of Borneo. The native name is "Rasong," and it is usually found near the mouths of rivers in the Southern part of Sarawak.

Hab. Sarawak (G. D. Haviland). Batang Lupar (C. Hose).

GENUS SEMNOPITHECUS.

SEMNOPITHECUS RUBICUNDUS.

Semnopithecus rubicundus, Müller, Tijdsh. voor natuur, Gesch. vol. v. pts. i. and ii., 1838, p. 137 (plate).

In the type of this species, all the animal is deep chestnut red, with the exception of the hands and feet, which are sullied with blackish. The hair on the frontal region is markedly radiated in all directions, that in front overshadowing the eyes, the eyebrows in the adults, wanting the remarkably long bristles which occur in them in the young. The hair on the vertex is laterally compressed, long semi-erect, being thrown somewhat backwards and tending to become recumbent. Whiskers but little developed. Face and ears bluish black; lips dull, flesh coloured. Nose depressed and slightly wrinkled. Tail concolorous with body, or darker, and tufted. Hair on the sides of body rather long. Under parts slightly paler than the upper. In the young, the general colour of the upper parts is purplish-red or brown; paler on the back, where it is mixed with yellowish hair, and still lighter on the head; the remaining parts being yellowish grey or white, except the tail, which is rather darker than the back.

	ft.	in.
Length of body to vent	2	0.00
Length of tail	2	5.75

This handsome red monkey is called by the Dyaks of Sarawak 'Jellu mirah,' and by the Kayans 'Kalasi,' and it is common everywhere. It is usually seen in large numbers, and some thirty or forty often pass one in the jungle, darting from branch to branch and making a tremendous noise. They will sometimes, when barked at by dogs, attack a dog and inflict a very bad bite. They ascend the mountains to the height of 3000-ft., but at that height the colour of their hair becomes a much deeper red. They are very destructive in fruit gardens.

Hab. Baram River (C. Hose). Mount Dulit (C. Hose).

SEMNOPITHECUS HOSEI.

Semnopithecus hosei, Thomas. P.Z.S. March 19th, 1889, p. 159—160. (Plate xvi.)

Size and form about as in *S femoralis*, *S. chrysomelas* and *S. obscurus*. Crown with a longitudinal crest starting about half-an-inch behind the centre of the forehead; the longer hairs slope evenly backwards, there being no trace of a reversed occipital tuft as there is in some species. General colour of body hoary grey, a colour made up by the intermixture of black and white hairs. Crest, centre of crown, and nape deep glossy black, as also are the long eyebrows, and the few short hairs scattered about the surface of the orbits. All the rest of the head, the forehead, temples, sides of the crown and neck, cheeks, lips, nasal septum chin (where there is a distinct tuft) and front of neck pure white, contrasting most markedly with the glossy black of the central cres, and with the dark grey of the back and shoulders. Outer sides of limbs like back, darkening terminally in the hands and feet to deep black. Chest, underside of body, and inner sides of limbs as far down as the middle of the forearm and of the lower leg white, continuous with that of the chin and throat. Tail hoary grey like the back throughout, only rather darker above than below, owing to the larger proportion of black as compared with white hairs there present.

Skull light and delicate. Nasal bones long, thin ; profile quite straight and continuous with the line of the forehead, an arrangement very different from the peculiar aquiline nasal outline of *S. femoralis*. Nasal opening oval, its breadth about two-thirds its height, instead of three-fourths as in the allied species. Bullæ low, opaque. Teeth as usual.

Dimensions of the type, an adult male, preserved in skin:

Head and body (c.) 520 millim.; tail 670; hind foot 154; heel to tip of hallux 123; length of eyebrows 25-28; length of crest-hairs (c.) 40.

Skull.—Greatest length (gnathion to occiput) 91 millim. basal length (basion to gnathion) 61 ; zygomatic breadth, 68; nasal opening, height 15.2, breadth 10.0; nasal length 10, greatest breadth 10; interorbital breadth 8.0; distance from outer edge of one orbit to that of the other 55.5; height of orbit, 23; breadth across the face, including external walls of orbits, 62; intertemporal constriction, 46; brain-case breadth 54, height from basilar suture to bregma (junction of sagittal and frontal sutures) 47; palate, length 30, breadth outside m1 30, inside m1 18.8; combined length of upper premolars and molars 26, of molars only 17.6. This handsome monkey is perhaps my finest discovery amongst the mammals. The type was shot at a place called Niah in the Baram district. I have since procured several specimens in different parts of the country, but although it is often seen in the low country I think we must consider it to be a mountain species which leaves the mountains at certain times in search of fruit. It ascends Mount Dulit to the height of 4000-ft., but is more common at 2000-ft. It frequents the salt-springs which are common in the interior, churning up the mud, and it is at these salt-springs that the Punans procure numbers of specimens with the blowpipe and poisoned arrows. From this monkey the Bezoar stones are obtained, being found either in the gall bladder or near it. The noise that this animal makes is loud and distinct—Gagah, gagah. The young resembles the colour of the adults, and are exceedingly pretty little things, but they won't live long in confinement and would never bear a voyage to England, as they suffer severely from sea-sickness. The Kayan name is 'Bangat.'

Hab. Niah (Type of species) (C. Hose). Baram River (C. Hose). Mount Dulit (C. Hose).

SEMNOPITHECUS FRONTATUS.

Semnopithecus frontatus, Müller, Tijdsch. voor Natuur. Gesch., vol. v., pl. i. and ii. 1838, p. 136.

Form slender; the face broad across the eyes, but compressed from above downwards; the trunk of the body dark yellowish brown, with a tinge of red on the flanks in some individuals, but passing into dark brown and then into black on the greater part of the outside of the limbs, on the back part of the thighs and on the root of the tail; the remainder of the tail being greyish or yellowish brown and tufted. The neck and head are yellow-brown, but the latter colour passes into black on the haired portion of the forehead, sides of the head, and on the crest. The under parts are pale reddish, lighter on the throat, extending as a narrow band down the inside of the forelimbs to near the wrist, and also on the hind limbs, but stopping short of the ankle. A bold, triangular area between the eyebrows, ascending to the middle of the forehead and in reality occupying the glabella, reaching at its upper end that part where the radiation of the hair of the other species usually takes place of a milky-white colour and wrinkled; the rest of the face being deep black, except the lower lip and a narrow line along the upper lip which are flesh-coloured and sparsely covered with yellowish brown hairs. Along the upper margin of the bare area are arranged long and black hairs, which being directed downwards and outwards commingle with the long bristly, black eyebrows, and reach as far back as the ears, as a marked lateral tuft or pencil, the hair on the side of the head above these lengthened hairs being short. The hairs of the cheeks from near the nose along the malar regions to the anterior root of the zygomatic arch are long and black, increasing in length on the hindmost part of the cheek to such a degree that they depend nearly to the shoulder. The crest is erect, high and compressed, occupying the middle line of the head like the ridge of a helmet, overreaching the forehead, where it is slightly contracted, and reaching backwards to the occiput, where it decreases and mingles with hairs on the upper part of the neck.

The young do not appear to differ much in colour from the adults.

	ft.	in.
Length of body to root of tail............	1	10.
Length of tail	2	4.

The skull of this remarkable animal is distinguished from the skulls of the other *Semnopitheci* by its highly arched, but rather narrow, retreating forehead ; by the great breadth of its orbits, the little naked prominence or arching of the interorbital space compared with that of the other crested species in which it is narrower than in *S. frontatus;* by its broad but short and truncated facial portion; and by the more retreating character and less depth of the symphysis of the lower jaw.

This species is very rare.

SEMNOPITHECUS FEMORALIS.

Semnopithecus femoralis, Horsfield. Appendix, Life Sir T. S. Raffles, 1830, p. 643.

This is a uniformly brownish-black monkey, the limbs, head, and tail being almost wholly black, but the difference between the colours is not well defined, and the fore limb is grizzled with whitish hairs. The tail is slightly bushy at its extremity. There is a rather short, vertical crest directed backwards, the hair anterior to it projecting forwards over the eyebrows. The ears are moderately large and partially exposed. The hair on the front and side of the head and on the middle of the crest is blackish or dull brown. The upper lip and chin are clad with short whitish hair, with longer black hairs intermixed on the chin. Along the flanks the hair is short, sparse, brown, and somewhat grizzled, which is the character also on the abdomen. The throat, sides of the neck and chest are concolorous with the upper parts. A narrow, well defined white line passes along the middle of the under surface from the chest in the adult to the hinder portion of the abdomen, but in the young specimens this line is obscure, as the throat, chest, and abdomen are yellowish white, and where it dies away on the inside of the limb, the white line is prolonged as a fine line to the wrist and ankle. In the adult the brachium is greyish, but there is a distinct tendency visible to the formation of a narrow obscure greyish line along the inner aspect of the antibrachium, but of variable intensity. The inner sides of the thighs are also white or pale grey, and this colour extends a short way below the knee. The face, ears, and the palms of the hands and sides of the feet black.

	ft.	in.
Length of body to root of tail	1	7.
Length of tail	1	10.

This monkey is a low country species seldom to be found on the mountains, and then only ascends to about 1000-ft. It is fond of living near the sea shore and is generally found in numbers of from ten to thirty sitting on the branches of tall trees in open spaces. Its Dyak name is 'Bigit,' that of the Kayans 'Pant.'

SEMNOPITHECUS EVERETTI.

Semnopithecus everetti, Thomas. P.Z.S. Nov. 1892, p. 582-583.

Mr. Thomas in his paper on this species makes the following remarks.

"In 1889 I had the pleasure of describing before this "Society a very beautiful species of *Semnopithecus* from the "Baram district, North-eastern Sarawak, which was discovered "by Mr. Charles Hose, and was named in his honour "*Semnopithecus hosei.* Of this monkey many specimens, all "from much the same district, have come to Europe, and I "have reason to believe that most of the European museums "have now been presented with examples of it, all obtained "by the same energetic and successful collector. In our own "museum we have, besides the type, another adult male, two "young specimens, and an adult skeleton. All these "specimens, including young ones barely a foot in length, "have shown the most striking uniformity in their coloration, "there being in none of them the smallest deviation from the "colour depicted in my illustration of the type (t.c. plate xvi.)

"Now, however, the museum has received, first from "Mr. A. Everett, who noticed the differences himself, one "specimen, and then from Mr. Hose two more, of a monkey "undoubtedly closely allied to *S. hosei*, but yet all three so "like each other, and so different in markings from any "specimen of that species which I have seen, that I feel "unable to consider it to be *S. hosei*, and therefore must "describe it as new.

"The chief difference lies in the distribution of the "colours of the head, for while in *S. hosei* only the centre of "the crown, and a narrow line down the nape and back, the "rest, including the whole of the region round and above the "ear, being pure white, in *S. everetti* the whole of the "forehead and top of the head are black, the lower limit of "the black passing across the middle of the ear, and the whole "breadth of the back of the neck is also black. A spot in the "exact centre of the forehead, just above the meeting of the "eyebrows is, however, pale yellowish white. The pale "cheeks and the pale sides of the neck are in this species in "just as striking contrast to the dark crown as in *S. hosei*, and "distinguish it equally from its near ally *S. chrysomelas (femoralis)*.

Mr. Thomas also says, "I must of course admit the "possibility of intermediate specimens between *S. hosei* and "*S. everetti* occurring, and the consequent necessity for the "reduction of the latter form to the rank of a sub-species; but "in the absence of such intermediate forms, and in view of "the great constancy in the coloration of *S. hosei* already "noted, it seems best to give a name to the striking variation "from it now described."

Since Mr. Thomas described this monkey I have obtained several other specimens of this species, and although I did not agree with Mr. Thomas in separating this from *Semnopithecus hosei*, I am bound to say that in every case the marking is quite constant. Both species are found on Mount Dulit, and also on Mount Batu Song, and as yet *Semnopithecus everetti* has not been found in the low country. The type was obtained by Mr. Everett on Kina Balu Mountain.

Hab. Mount Dulit (C. Hose). Mount Batu Song (C. Hose). Kina Balu (Type of species) (A. Everett).

SEMNOPITHECUS CRISTATUS.

Semnopithecus cristatus, Müller, Tijdschr. voor Natuur. Gesch. en. Phys. vol. ii. 1835, p.p. 316, 328.

The young are reddish fawn, but the hands and feet gradually change through greyish brown to the colour of the adult, the crest also with increasing age becoming directed forwards.

Length of body to the vent, 2ft. 2in.
Length of tail 2ft. 6in.

These silver grey monkeys are fairly common in the low country, and are called by the Dyak, 'Bigok,' and by the Kayans, 'Chikok,' from the noise they make. They ascend the mountains to about 2,000 feet.

Hab. Baram River (C. Hose). Mount Dulit, 2,000 feet, (C. Hose).

SEMNOPITHECUS CRUCIGER.

Semnopithecus cruciger, Thomas, Ann. and Mag. Nat. Hist. Ser 6, vol. x. Dec. 1892.

This is a most remarkable monkey which has lately been described by Mr. Thomas from a flat skin obtained by me some years ago at a place called Miri in the Baram district. I always considered the skin to be merely a striking variety of *Semnopithecus femoralis*. I shot the monkey on the sea coast along with a number of *Semnopithecus femoralis*; this was in the year 1887. I afterwards had the skin of a baby brought in, the marking of which were similar to that of the type, and it was obtained within a few miles of Miri, at a place called Bakam. But in September, 1892, one of my Dyak hunters procured three fine adult specimens of this monkey in the Batang Lupar river in Southern Sarawak, and reported that they had seen several other specimens of like marking. In the three adult specimens the black cross down the centre of the back in some cases is broken, and the thighs are darker in some cases in one than the other, but the st iking red marking is kept up throughout each specimen.

Fur long and soft on the head and shoulders. Hairs of crown especially long, standing vertically upright everywhere, so that there are no centres of convergence or divergence, but that along the median line is somewhat longer than that on the sides, and there is therefore an ill-defined crest. Colours of crown, sides of body from axilla, haunches, and outer sides of legs to ankles, brilliant red, rather more chestnut on the head and paler on the lower legs. Hands, outer sides of arms to shoulders, nape and central line down the back from the withers on to the base of the tail deep glossy black. In some cases the black line down the back is broken with a mixture of red and black hairs.

Eyebrows black, contrasting markedly with the red forehead; short hairs of face, whiskers, hairs of ears, sides of neck, whole of chin, chest and belly, and lines down inner sides of arms to wrists, and legs to ankles glossy white, with a faint yellowish suffusion.

Tail above black basally, gradually becoming duller at the tip.

Type of species Miri River, Sarawak (C. Hose). Batang Lupar River (C. Hose). Bakam River (C. Hose).

ORDER LEMUROIDEA.

FAMILY LEMURIDÆ.

GENUS NYCTICEBUS.

NYCTICEBUS TARDIGRADUS.

Nycticebus tardigradus, L. Syst. Nat. page 44 (1766).

Fur close and woolly, covering the whole of the face and body with the exception of the nose and lips. The short hairy ears and the short tail are almost concealed beneath the fur. As a rule there are four incisors in the upper jaw, but one or both of the outer pair may be wanting. Each eye is surrounded by a dark brown circle, broadest above; nose and soles of the feet flesh colour where naked. Habits purely nocturnal and arboreal. This animal feeds on leaves and shoots of trees, fruits, insects, bird's eggs, and young birds. It sleeps rolled up in a ball, its head and hands buried between its thighs, and wakes up in the dusk of evening to commence its nocturnal rambles. The female bears but one young one at a time. The Dyak name is 'Kukong,' and the Kayans call it 'Bengkong.' It is a low country species.

Hab. Baram River (C. Hose). Niah River (C. Hose).

FAMILY TARSIIDÆ.

GENUS TARSIUS.

TARSIUS SPECTRUM.

Lemur spectrum, Pallas, nov. spec., Quad. e Glir. ord. 1778, p 275 nt.

This curious little animal is found in the jungle of the low country skipping about from branch to branch. It has a habit of turning its head almost completely round without moving the other part of its body. It has very large eyes and curious pads on each of its fingers. Dyak name 'Ingkat.'

Hab. Baram River (C. Hose). Lundu River (A. Everett).

ORDER CARNIVORA.

FAMILY FELIDÆ.

GENUS FELIS.

FELIS NEBULOSA.

Felis nebulosa, Griffith, p 37, plate (1821).

The clouded leopard has the tail thickly furred, nearly the same thickness throughout, and long, about four fifths the length of the head and body. Caudal vertebrae 25.

Skull long, low, narrow. Orbit widely open behind. Hinder termination of bony palate concave; mesopterygoid fossa narrow. Lower edge of mandible straight from symphysis to near the angle, then concave. The upper canines are longer relatively than in any other living cat, and have a very sharp edge posteriorly. Anterior upper premolar frequently but not always wanting. Colour, general tint varying from greyish or earthy brown (cat-grey) to fulvous (light yellowish brown); lower parts and inner sides of limbs white or pale tawny. Head spotted above; two broad black bands, with narrower band or elongated spots between them commence between the ears, run back to the shoulders, and are prolonged, more or less regularly, as bands of large oval or elongated marks along the back. Sides of the body usually divided into large subovate, trapezoidal, or irregular shaped darker patches by narrow pale bands, the patches in places edged with black, especially behind. In old specimens the dark patches are sometimes indistinguishable, but the black edges remain as irregular stripes. The limbs and under parts are marked with large black spots. Tail with numerous dusky rings, often interrupted at the sides, those near the body traversed above by a longitudinal band.

Measurements—37½ inches long from snout to vent; tail with hair at end 30, without 20; height 14½; length of ear 2⅜; weight 44½-lbs.

This animal is constantly procured by the natives of Borneo, the canine teeth being used by the Kayans and Kenniahs as ear ornaments, and the skin for the purpose of a war coat. It is found both in the low country and on the mountains to a height of 5000-ft. Dyak name 'Enkali.' Kayan name 'Kolih.'

Baram River (C. Hose). Mount Dulit 5000-ft. (C. Hose). Batu Song 2000-ft. (C. Hose).

FELIS MARMORATA.

Felis marmorata, Martin, P.Z.S. 1836, p. 108.

This marbled cat is larger than a domestic cat. Tail bushy, nearly the same thickness throughout, about three quarters the length of the head and body. Fur soft, thick, with woolly underfur. Ears short, rounded at the end. Bony orbit complete behind in old skulls. The posterior edge of the bony palate deeply concave. Anterior upper premolar apparently often wanting. Ground colour varying from brownish grey (earthy brown) to bright yellowish or rufous brown, lower parts paler. The sides divided by narrow pale streaks into large, irregularly shaped darker patches, black on the hinder edges. Along the back are angular black blotches or irregular rings, arranged more or less in longitudinal bands. There are black spots on the outside of the limbs, the upper surface of the tail, and usually on the lower parts; but those on the belly are very variable, being sometimes large and distinct, sometimes almost imperceptible. The inside of the limbs and the chest are banded or spotted, and there are the usual cheek stripes. Two interrupted bands, one from the inner corner of each eye over the head, are continued as well marked black stripes on the neck, spots or bands intervening between them on the head, but not on the neck. The under fur is rich brown. According to Blyth, the ground-colour becomes more fulvous with age.

Dimensions—Length of head and body, 18½ inches to 23 inches; tail, 14 to 15½ inches; ears from crown of head, (2 Jerdon). The basal length of skull is 2·95 inches, zygomatic breadth 2·6.

The habits of this animal are similar to those of *Felis nebulosa*, but it often has been known to frequent the clearings, and is more often found in the low country than on the mountains. It is very fierce when caught, and will not live long in captivity.

Hab. Baram River (C. Hose). North Borneo (A. Everett).

FELIS BENGALENSIS.

Felis bengalensis, Kerr, Animal Kingdom, p. 151 (1792); Blyth, P.Z.S. 1863, p. 184.

This animal is about the size of a domestic cat or rather smaller, but with longer legs. Tail rather less than half the length of the head and body together, sometimes perhaps not more than one third, but some measurements give more than one half. Ears moderate, rounded at the tip. Pupil circular (perhaps elliptical in strong light). The skull is rather elongated, low and convex, orbit incomplete behind. The inner lobe of the upper flesh-tooth small. Anterior upper premolar rarely deficient.

Colour, ground colour above pale fulvous, varying from rufous to greyish, below white, ornamented throughout with numerous more or less elongate, well defined spots, either black throughout, or especially on the sides, each spot partly black and partly brown, the two colours passing into each other. The fur is brown at the base, and many of the fulvous hairs have white tips, producing a grizzled appearance on the ground colour. The sizes of the spots are very variable. Dimensions: head and body 24 to 26 inches, tail 11 to 12 or more.

This pretty little cat is found in the low country, and on the mountains to the height of 3,000 feet. It is constantly trapped by the natives, and it is very fond of stealing fowls, going into the villages and taking chickens from beneath the houses. The Dyak name is 'Kuching Batu.' It usually lives amongst the rocks and in holes of trees.

FELIS TEMMINCKI.

Felis temmincki, Vigors & Horsf. Zool. Journ. III, p. 451 (1828);

Size rather less than that of *F. nebulosa*. Tail about two thirds the length of the head and body, almost the same thickness throughout. Caudal vertebra 22. Ears short, roûnded. Fur moderate length, dense, rather harsh. Colour, deep ferruginous or chestnut, darker (bay) along the back, paler on the sides, still paler and whitish below; chin and lower surface of tail to the tip white, the tip above is dusky. There are some round dusky spots on the breast, between and behind the axils, and, in some specimens, on the inside of the fore limbs, and less distinct markings, forming imperfect bands, on the throat. The lower side of the tarsi and feet are brown. The markings on the face are peculiar and somewhat variable; the most conspicuous is a ho izontal white or buff cheek stripe, sometimes edged with black, from below the eye to behind the gape; a whitish band inside each eye; and occasionally curved lines running back from above the eye to between the ears. Ears black or brownish black outside, with an ill-defined pale central spot. Fur brown at the base, ferruginous near the end, some black tips on the back.

This cat is very rare in Borneo, as yet I have not met with it in the Baram district. Dimensions: length of head and body 31·5 inches, tail 19, height at the shoulder 17, length of ear, 2·5.

FELIS PLANICEPS.

Felis planiceps, Vigors and Horsf.

About the size of a domestic cat. Tail short, a quarter to a third the length of the head and body. Orbits completely enclosed by bone, and the anterior upper premolar larger and better developed than in any other living cat, having two roots. Colour, dark rich brown above, the fur having a silvery speckled appearance, owing to an intermixture of hairs with white tips; below white, more or less spotted or splashed with brown.

This cat is common in the low country, and is often very destructive in the gardens. It is very fond of fruit, and has constantly been known to dig up and eat the sweet potatoes which are grown by the natives of Borneo. Dyak name, 'Jellu Maio'; Kayan, 'Using.'

Hab. Baram River (C. Hose). Mount Dulit 2000-ft. (C. Hose).

FELIS BADIA.

Felis badia. Gray, P.Z.S., 1874, p. 322, pl. xlix.

This handsome red cat is very rare, and only met with in the dense forest. It is about the size of *Felis marmorata*, but the general colour is a dark chestnut red. I have not had an opportunity to notice the habits of this animal, having only obtained one specimen.

Hab. Suai River, Sarawak (C. Hose).

FAMILY VIVERRIDÆ.

GENUS VIVERRA.

VIVERRA TANGALUNGA.

Viverra tangalunga, Gray, P.Z.S. 1832, p. 63;

This Civet cat inhabits the low country of Borneo, and ascends the mountains to the height of 3,000 feet. It has a very musky smell, and is called by the Dyaks, 'Sinang,' and the Kayans, 'Tangalang.' The body measures two feet from the tip of the nose to the root of the tail. The species is figured in the P.Z.S. for 1876 (plate xxxvii.), and it is very common in Borneo.

Hab. Baram River (C. Hose). Mount Dulit, 3,000 feet (C Hose). Sarawak (A. Everett). Mount Kalulong (C. Hose).

GENUS LINSANG.

LINSANG GRACILIS.

Felis gracilis, Horsf. Zool. Researches, 1824.

This animal is strikingly characterised by a slender body, a tapering head and sharp muzzle, a long and thick tail, and slender delicate limbs. The body in length is nearly equal to that of the domestic cat, but, in consequence of its slender make, it has a greater resemblance in form to the various species of Viverra. Dimensions: length of body from nose to root of tail, 1ft. 3½in.; the head, 3½in.; the tail, 1ft. ½in.

On a ground of pale yellowish white, which covers the throat, breast, belly, sides, and part of the back and tail, the distinguishing marks of a deep brown colour, inclining to black, are arranged in the following manner: four transverse bands, gradually increasing in breadth, cover the back at intervals between the limbs; on the rump are two narrow bands; two longitudinal stripes take their origin, one between the ears, the other near the posterior angle of the eye, on each side, and pass, with interruption at the transverse bands, to the thighs, when they are continued by numerous large spots which cover these parts.

Hab. Baram River (C. Hose).

GENUS PARADOXURUS.

PARADOXURUS HERMAPHRODITUS.

Paradoxurus hermaphroditus, Gray, P.Z.S. 1832, p. 67; 1859, p. 113.

Tail more than three quarters the length of the head and body, sometimes quite as long or a little longer. Fur of moderate length in general, not so long and ragged as in *P. niger*. Colour, brownish grey, occasionally ashy. Under fur, when present, brownish, the longer hairs light brown or grey, with occasionally black tips, but these are not as a rule greatly developed, though some Bornean specimen have long sooty brown termination. The back is generally more or less distinctly striped with black longitudinally, the number of stripes varying, and the lateral bands, being often replaced by rows of spots. The markings are very variable, and in some cases I have obtained specimens with white tips to the tails. This species is very common both on the mountains and on the low country. and is very destructive to poultry. The Dyak name is 'Munsang.'

Mount Dulit, 2,000 feet (C. Hose). Mount Lambir, 1,000 feet (C. Hose). Sarawak (A. Everett).

PARADOXURUS LEUCOMYSTAX.

Paradoxurus leucomystax, Gray, P.Z.S. 1836, p. 88.

This is the largest species of the genus except *P. musschenbroecki*. Tail about three quarters the length of the head and body. Colour in general rufescent brown, paler below; under fur, greyish yellow at the base, the longer hair bright rufous, and the tips dusky or black. Sometimes this animal has a white tip to the tail.

Dimensions: head and body 27 inches; tail 20 inches; skull about 5 inches long. Dyak name, 'Galling.'

Hab. Mount Dulit, 3,000 feet (C. Hose). Sarawak (G. D. Haviland).

GENUS ARCTOGALIDIA

ARCTOGALE LEUCOTIS.

Paradoxurus leucotis, Blyth, Horsf. Cat. p. 66 (1851);

Tail about the same length as the head and body. Fur short, of uniform length, not harsh. Skull narrow and elongate. Post orbital processes long, zygomatic arches weak. The bony palate extends more than half an inch behind the last upper molars. Colour, fulvous grey (whity brown) to dusky grey, or occasionally brown above, much paler below. Fur in pale specimen sometimes grey throughout; in darker skins brown near the base, then grey, tipped on the back with dark brown or black. Along the back run three longitudinal dark bands, either continuous or broken into spots; sometimes in the Bornean specimens the bands are indistinct or wanting. The head above, including the crown and ears, usually darker, often ashy or black.

Dimensions: head and body of a large male 26·5 inches; tail 27; skull 4-in. in basal length, 2·3 in zygomatic breadth.

This mammal is frequently met with in the low country; as yet I have not obtained a specimen from any of the mountains in Borneo. The Dyaks call it 'Musang akar.' It is usually seen running along the branches of trees in the old forest.

Baram River (C. Hose). Bakong River (C. Hose). Mount Kalulong (C. Hose).

GENUS HEMIGALE.

HEMIGALE HARDWICKEI.

Hemigale hardwickei, Gray, Spic. Zool. ii. p. 9, t. 1.

The skull agrees with *Genetta* and *Nandinia* in the hinder opening of the palate being only a short distance behind the line between the back edges of the hinder tubercular grinders. The nose of the skull is elongate. The brain cavity ovate, ventricose, not suddenly constricted in front.

Length of skull, 3·75 inches. The coloration is very peculiar, pale brownish grey, with a variable number (usually 5 or 6) of broad, dark transverse bands on the back, longitudinal stripes on the nape, and rings on the basal portion of the tail.

Found in the low country and on the mountains to the height of 2,000 feet. It is easily caught by the natives in spring traps called 'jirat.' The Dyak name is 'Pankat tekalang,' but the Kayan name is 'Tekalang alud.' The meaning of alud is a boat, and the reason for the name is because the striking marking on the back resembles the seats of a boat.

Hab. Baram River (C. Hose). Sarawak (A. Everett). Batu Song (C. Hose). Mount Dulit 2,000-ft. (C. Hose).

HEMIGALE HOSEI.

Hemigale hosei, Thomas Ann. Mag., N. H. (6) ix. p.p., 250. (1892). P.Z.S. 1892. p. 222. pl. xviii.

Size and proportions very much those of *H. hardwickei*, although the skull seems to be rather more lightly built. General colour above uniform dark smoky brown or black, the bases of the body-hairs whitish. Sides of muzzle at the roots of the whiskers white, the corresponding place in *H. hardwickei* being black; cheek below eye and a patch above and behind it grizzled brownish white. Ears thinly haired, pure white on their inner aspect; edges in marked contrast to the black crown. Chin white: chest, belly, and inner sides of limbs proximally smoky yellowish grey. Rest of limbs and whole of tail black.

Skull (P.Z.S. 1892 Plate xix. figs 1—3) rather slenderer and lighter than that of specimens of *H. hardwickei* of similar age and sex, muzzle rather more parallel-sided, not tapering so much anteriorly. Infraorbital foramina comparatively large. Dimensions—Head and body 540 millim; tail, 320; hind foot, 78; skull, basal length 89; greatest breadth, 45·3; interorbital breadth, 18·8; tip to tip of postorbital processes, 22·5; intertemporal breadth, 14·3; palate length 54; breadth at posterior corner of $\overline{p4}$ 25; length of palatine foramina, 5·2; greatest diameter of infraorbital foramina, 5·9.

Mr. Thomas remarks in his paper on the mammals of Mount Dulit, which were sent to the Museum by me in 1892, "This striking species is certainly the chief prize of the "collection, as new Carnivores are very rare, and so distinct a "new species has not been described for many years." The habits of this mammal are like those of *H. hardwickei*, but it is a true mountain species, as it is only found between the heights of 2000 and 4000-ft.

Hab. (Type of species) Mount Dulit 4000-ft. (C. Hose). Mount Batu Song 2000-ft. (C. Hose).

GENUS ARCTICTIS.

ARCTICTIS BINTURONG.

Arctictis binturong, Temm. Mon. Mamm., ii. p. 308.

Tail nearly as long as the head and body, very thick at the base, clothed with bristly, long, straggling hairs, longer than those of the body. Fur coarse and long, some piles longer than the rest of the fur, especially on the back. Colour black, more or less grizzled on the head and outside of the fore limbs, and sometimes throughout the body. On the head and outside of the fore limbs, and often on the back, there is a subterminal grey or rufous-grey ring on the longer hairs. In young specimens there are long grey or rufous tips to the fur. The ears have a white border, but the tufts are black. Dimensions—Head and body, 28 to 33 inches; tail, 26 to 27.

This animal is omnivorous, living on small mammals, birds, fishes, earth worms, insects, and fruits; it is also nocturnal and arboreal, its power of climbing about trees being much aided by its prehensile tail. These animals are common in parts of Borneo, usually living in the dense forest, but when in search of fruit they will often visit gardens. They become very tame in captivity, and are called by all natives 'Binturong.'

Hab. Baram River (C. Hose). Sarawak (A. Everett).

GENUS CYNOGALE.

CYNOGALE BENNETTI.

Cynogale bennetti, Gray, Proc. Zool. Soc. 1836, p. 88.

This is a remarkable mammal, somewhat resembling an otter in form. It is grizzled grey on the head and back, which becomes a brownish colour towards the hind quarters, the grey disappearing altogether in the tail. Its legs are brownish black, and the belly a lighter shade. The feet are webbed. Chin, dirty white; long, hard, white hairs coming from the upper lips and cheeks, with a few black ones intermingled; tail short. This mammal is found on the banks of rivers in the low country, and on the mountains. I also once met with a specimen on the sea coast; it is particularly fond of living near swampy places. Its food probably consists of small fish and frogs. When it is being chased by dogs it prefers to seek refuge in climbing the trees rather than taking to the water. I once obtained a specimen in a fish trap. The Dyak name is 'Gelu labbi,' and also 'Paddy barhu,' because it smells like new paddy.

Hab. Baram River (C. Hose), Lawas Mountains (A. Everett), Marudi Hills (E. Cox).

GENUS HERPESTES.

HERPESTES BRACHYURUS.

Herpestes brachyurus, Gray, Mag. Nat. Hist. (new series), i. 1837, p. 578.

The type of this species is in the British Museum. The general colour is dark blackish brown, finely punctulated with yellow, more especially on the anterior half of the body and on the shoulders, the caudal hairs being broadly black tipped, and the head paler and more olive brown than the rest of the body. The fore limbs and the lower half of the hind legs are dark brown and unspeckled. The chin and throat are rusty yellowish brown; the chest and belly are brown, and the hairs are banded much as on the back. The tail is untufted, and it is broader at its base, from which it gradually tapers to the tip. Length from tip of muzzle to root of tail, 17·50 inches. Length of tail without hairs 7 inches. Cantor records a male 18·50 inches long, with the tail 9 inches. This mongoose is fairly common in Borneo, being found all through the low country, and on the mountains to the height of 3,000 feet. It has a very peculiar smell. The Dyak name is 'Dumbang'; Kayan name, 'Dubang.'

Hab. Baram River (C. Hose), Mount Dulit, 3,000-ft., (C. Hose), Suai River (C. Hose), Sarawak (A. Everett).

HERPESTES SEMITORQUATUS.

Herpestes semitorquatus, Gray Ann. and Mag. Nat. Hist. 1846, vol. xviii., p. 211.

This species is easily distinguished by the pale area along the side of the neck, from whence it derives its specific name. The general colour of the animal is rich orange-brown, most intensely rufous on the sides of the body, the back and upper parts of the side being finely marked with yellow, which becomes very indistinct on the shoulders and outside of the thighs; the fore legs and the lower half of the hind legs are dark purplish-brown. The lower half of the sides of the neck from the extremity of the muzzle backwards below the ear to the front of the shoulder, is a rufous yellow and clearly marked off from the colour of the upper part of the neck, which is dark rufous-brown and punctulated, while the underlying neck-band is not.

Dimensions—Length of head and trunk, 17·30 inches; tail without hair, 10·50 inches; tail with hair, 11·70 inches.

The only skull seen of this species is distinguished by its rather broad muzzle, and by apparently an imperfect orbit, as all the sutures are lost, and yet the two processes are far apart. It is found on the mountains to the height of 3500-ft., where it is more common than in the low country. It is more often seen amongst the large boulders and rock on the slopes of the mountains than in the dense jungle. The Dyak name is 'Dumbang mirah.'

Hab. Baram River (C. Hose). Mount Dulit 4000-ft. (C. Hose). Mount Batu Song 3000-ft. (C. Hose).

FAMILY CANIDÆ.

GENUS CYON.

CYON RUTILANS.

Cyon rutilans, S. Müll. Verhandelingen, Zool. Zoogd., p.p., 27, 51 (1839).

Colour, uniform deep ferruginous red above, hair scarcely paler towards the base. Lower parts whitish. Terminal portion of tail black.

Dimensions—Head and body, 32½ inches in a young male; tail, 12; tarsus and hind foot in adults, 6 inches.

This wild dog must be very rare in Borneo. I have constantly heard native accounts of it, but I have never seen a specimen. The natives state that these dogs hunt wild animals in packs. They have many superstitions concerning these animals, and they are spoken of as a "hantu" or spirit. Dyak name, 'Pasun.'

FAMILY MUSTELIDÆ.

GENUS MUSTELA.

MUSTELA FLAVIGULA.

Mustela flavigula, Bodd. Elench. An., p. 88 (1785).

Tail long and bushy, measuring, without hair, quite three quarters the length of the head and body. Caudal vertebræ 24. Feet more or less naked beneath; in Malay specimens the whole metacarpus and more than half the tarsus are bare, whilst in some Himalayan animals the naked soles appear less developed.

Colour glossy blackish brown; chin, upper part of throat as far as below the ears, white; throat and breast yellow or orange or brownish yellow.

Dimensions—Head and body, 22 inches; tail, without hairs, about 16; with hairs, 17 to 20.

This animal is found both in the low country and on the mountains to the height of 2000-ft. It constantly frequents the fruit gardens of the natives, and lives in the holes of hollow trees during the greater part of the day time, coming out to feed about three hours before sundown.

Hab. Baram River (C. Hose). Mount Dulit 2000-ft. (C. Hose). Mount Kalulong (C. Hose).

GENUS PUTORIUS.

PUTORIUS NUDIPES.

Putorius nudipes, F. Cuv. (Hist. Nat. Mamm. pl. 140);

Tail bushy. Soles partly naked. Fur loose and long, with but little underfur. Colour rusty red, the head above and below white, tail-tip whitish. Head and body about 13 inches, tail without hair 8½, with hair 10½; skull 2·25 inches long, 1·35 broad.

This animal is very rare in the Baram district, I have met with but one specimen on Mount Kalulong, but it is more common near Kuching, where Dr. Haviland obtained several fine specimens.

Kuching (G. D. Haviland). Mount Kalulong (C. Hose). Sibuti River (E. Cox).

GENUS MYDAUS.

MYDAUS MELICEPS.

Mydaus meliceps, F. Cuv. Hist. Nat. des Mammifères (1821).

This curious skunk-like animal is found in the northern part of Borneo. Back, black; crown, white, reaching to the middle of back; tail short, snout long, claws on the fore legs large.

Hab. North Borneo (A Everett).

GENUS LUTRA.

LUTRA SUMATRANA.

Barangia sumatrana, Gray, P.Z.S. 1865, p. 123.

This is a large otter, the length of the head and body in an old male, according to Cantor, being 32½ inches; tail 20, and the colour deep rich brown throughout, except on the chin and throat which are whitish. The nose is entirely hairy in the young specimens, but in the older individuals the hairs become partially worn off. This otter is found along the banks of the rivers, and occasionally on the sea coast. It is called by the Dyaks, 'Ringin,' and the Kayans, 'Dingin.' The Malays sometimes call it the 'Anging ayer,' or water dog.

Hab. Baram Mouth (C. Hose). North Borneo (A. Everett).

LUTRA CINEREA.

Lutra cinerea, Illiger (1815).

Mr. Thomas shows that the clawless otter, *Lutra leptonyx*, of Horsfield (1824), must take the earlier title of *L. cinerea*, Illiger. This clawless otter is very rare in Borneo. It is called by the Malays, 'Nabrang.'

Hab. Kuching, Sarawak (J. E. A. Lewis).

FAMILY URSIDÆ.

GENUS URSUS.

URSUS MALAYANUS.

Ursus malayanus, Raffles, Tr. Linn. Soc. xiii., p. 254 (1822).

Size small. Fur short and coarse. Claws well curved. Ears small, rounded, covered with short hair. Tongue very long. Skull in adults very short and broad, nose short, zygomatic arches wide. Auditory bullæ more swollen than in *U. arctus* or *U. torquatus*. Incisor and canines large, premolars crowded and soon lost. Upper sectorial very small, its transverse section scarcely larger than that of the outer incisor. Molars short and very broad.

Colour—Black, brownish in parts. The muzzle, including the eyes and the chin, paler, often whitish; the crescentic patch on the chest white, yellow, or orange, with the two end often broad, sometimes united into a large oval or heart-shaped spot with a black centre, and sometimes with the apex prolonged into a white streak on the abdomen. Claws pale horny, sometimes dusky.

Dimensions—Head and body, about 4-ft.; tail, 2 inches; hind foot 7 inches.

This little Bornean bear is fairly common both in the low country and on the mountains to the height of 3000-ft. It is very fond of feeding upon the honey of a very small bee called by the Dyak 'Kalulut,' and I have seen holes in trees of hard wood made by the bear with its claws in its endeavours to get at a nest of these bees. The bees usually have but a very small hole for an entrance. Bears are not eaten by the natives of Borneo as a rule, as people who eat bears' flesh are supposed to go mad. The skins are often used for war coats, and the gall bladder ('impadu') is readily bought by the Chinese traders from the natives for medicinal purposes. The Dyaks call a bear 'Jugum,' the Kayans 'Buang,' and the Malays 'Bruang.'

Hab. Baram River (C. Hose). Suai River (E. Cox). Mount Dulit 2000-ft. (C. Hose). Ridan River (E. Cox).

ORDER INSECTIVORA.

FAMILY TUPAIIDÆ.

GENUS TUPAIA.

TUPAIA JAVANICA.

Tupaia javanica, Horsfd. Zool. Resch. in Java, 1822, fig.

Dr. Günther in his paper on Bornean Mammal in the P.Z.S., May 16th, 1876, p. 426, remarks that "Distinct as this species is from *Tupaia ferruginea*, young examples of the latter can hardly be externally distinguished from *Tupaia javanica*; and one of the specimens examined and named *Tupaia javanica* by Horsfield is clearly the young of the larger species, as it is proved by its undeveloped dentation."

I have never met with *Tupaia javanica* in Borneo. and I think perhaps it is possible that specimens considered to be *javanica* from Borneo, are merely forms of *Tupaia minor*, which is very common in all parts of the Island.

TUPAIA FERRUGINEA LONGIPES.

Tupaia ferruginea longipes, Thomas, Ann. and Magazine of Natural History, Ser. 6, vol. xi., May 1893.

Hind feet conspicuously longer than in the typical Sumatran form. General colour less ferruginous above, but more so below; the shoulder-streak also, instead of being yellowish or whitish, is rich rufous. Upper-side of tail concolorous with the back, instead of being markedly greyer.

Skull and teeth apparently as in the typical sub-species.

Dimensions of type (a skin):—Head and body, 192 millim.; tail, 190; hind foot, 45·5.

This tree shrew is a long footed form of *Tupaia ferruginea*, examples of which have constantly been found in Borneo by Mr. Low, Mr. Everett, and myself. The true Sumatran form of *Tupaia ferruginea* to the best of my knowledge has not been met with in Borneo, its place being taken by this longfooted sub-species. It occurs both on the mountains and in the low country, ascending the mountains only to the height of 2000-ft. It is rare and very difficult to see in the jungle on account of its colour.

Hab. Rijang River (H. B. Low), Batu Song (A. Everett), Mount Dulit, 2000-ft. (C. Hose), Ridan River (C. Hose), Baram River (C. Hose).

TUPAIA TANA.

Tupaia tana, Raff. Trans. Linn. Soc. vol. xiii. 1821, p. 257;

Fur moderately long and fine. It consists of two kinds of hairs—long, entirely black; rather stiff hairs, and shorter hair with a sub-apical orange or dark rufous-brown band. The former kind occurs most numerously on the interscapular black band and on the hind quarters; the orange-banded hairs cover the head, where they are very short the shoulder band, and a rather broad area below the interscapular band. The dark rufous-brown hairs occur chiefly on the sides, the colour of which and of the limbs is a deep ferruginous chestnut, gradually passing into black on the back. The tail is dark, very deep ferruginous-chestnut and bright rusty on the under surface. The chin and the throat are rusty brown, the chest and belly being paler chestnut than the upper parts.

The skull is at once distinguished from the skulls of all known *Tupaiæ* (except *Tupaia everetti* of the Philippines) by the long attenuated character of the pre-ocular portion, and by the length of the snout.

This tree shrew is one of the most common in Borneo, it is found all through the low country, and on the mountains to the height of 4000-ft.

Hab. Mount Penrisen (A. Everett), Mount Dulit 4000-ft. (C. Hose). Batu Song Mount 2000-ft. (A. Everett).

TUPAIA MINOR.

Tupaia minor, Günther, P.Z.S. 1876. p. 426.

All the hairs of the upper parts are grizzled with grey, brownish grey, and black, a reddish brown tinge prevailing in the middle of the hinder half of the back and on the tail; extremity of the tail black. Shoulder-stripe distinct. Lower parts yellowish white, of the tail brownish yellow. With the exception of the terminal hairs, the hairs of the tail are rather short.

Length of body 5 inches 4 lines, of tail 6 inches 2 lines.

This little tree shrew is fairly common both on the mountains to the height of 4000-ft. and in the low country. It breeds in a nest in an old stump covered with creepers, but I am not sure whether it makes the nest itself or occupies the nest of a bird. I have found two of these nests, but the material used was different.

The Dyak name for all the Tupaia family is "Tupai Tanah" which means ground squirrel.

Hab. Baram River (C. Hose), Limbang River (C. Hose). Sarawak (A. Everett). Mount Dulit 3000-ft. (C. Hose), Mount Batu Song (C. Hose).

TUPAIA DORSALIS.

Tupaia dorsalis, Schleg.

This tree shrew is easily distinguished from any of the other Tupaiæ by the very distinct black line down the back.

Hab. Baram River (C. Hose). Sarawak (A. Everett). Mount Dulit, 4,000-ft. (C. Hose). Mount Kalulong (C. Hose). Mount Batu Song (C. Hose).

TUPAIA SPLENDIDULA.

Tupaia splendidula, Gray, P.Z.S., 1865, p. 322, pl. xii.

The tail of this animal is less than the length of the body and head, and is blackish chestnut, and in strong contrast to the tail of *Tupaia ferruginea*, which is, so to speak, of a blackish olive, while the tail of *Tupaia tana* may either be rich chestnut rump. Dr. Gray describes the species as follows :—

"Fur, dark brown, blackish washed. Tail, dark red-"brown, pale red beneath, longer than the body and head; the "shoulder streak yellow; no bands between the shoulders. "The head conical, about twice as long as wide behind."

The head is large, compared with th· size of the body; the ears rounded, with several ridges on the conch, and a well developed convex tragus, not unlike the human ear.

Hab. (type of species) (Low), Borneo.

TUPAIA PICTA.

Tupaia picta, Thomas. Ann. Mag. N.H. (6) ix. p. 251, 1892.

Rather smaller than *Tupaia ferruginea*; more heavily built than *T. dorsalis*. General colour of back, olive grey, coarsely grizzled with yellowish; more rufous posteriorly. Centre of back with a distinct dorsal stripe extending from the withers to the rump, the stripe better defined than in *Tupaia montana*, but neither so long nor so sharply defined as in *Tupaia dorsalis*. Head, hands and feet, dull grizzled olive; sides, dark rufous; a distinct shoulder stripe present. Under side grey, the hairs washed terminally with yellow; chin and chest, rich yellow or orange. Tail, broad and bushy, evenly distichous, its hairs above basally mixed red and black, at the tip and below, brilliant chestnut rufous. Skull much as in *Tupaia ferruginea*, but the zygomatic vacuity reduced to a long narrow slit about 4·5 millim. long, and only about 1 millim. high. Teeth also not materially different from those of *Tupaia ferruginea*.

Dimensions of type.—Head and body, 185 millim; tail, 162; hind foot, 42·5. Skull: basal length, 45; greatest breadth, 26·3; nasal tip to front edge of orbit, 21.

This handsome species is readily distinguished from *Tupaia ferruginea* and *Tupaia splendidula* by its duller body colour, and the presence of a black dorsal stripe; from *Tupaia tana* by its smaller size and shorter muzzle; from *Tupaia dorsalis* by its less defined line, bushier tail and heavier teeth; and from *Tupaia montana*, described above, by its brilliantly rufous tail and coarsely grizzled back.

This tree shrew is more common in the low country than on the mountains; it is usually found in the dense forest, and particularly active in its movements.

Hab. Baram River (C. Hose). Mount Dulit, 2,000-ft. (C. Hose). Batu Song Mount (C. Hose). Ridan River (type of species) (C. Hose).

TUPAIA MONTANA.

Tupaia montana, Thomas, P.Z.S. March, 1892, p. 223-224.

Size much as in Malaccan specimens of *Tupaia ferruginea* (Bornean ones are rather larger), but the tail shorter in proportion. General colour above, dusky olive, with a strong rufous suffusion; head clearer olive. Back, in fully adult specimens, with a deep black median line running from the withers to the rump, but broadening out, and becoming less sharply defined in its posterior half. Under surface, greyish orange, the hairs grey at their bases, broadly washed terminally with rich olive yellow. Tail concolorous with the body, not greyer, as it is so markedly in *Tupaia ferruginea*; grizzled black and shining ferrugineous above; below the central short-haired part is grey; then, laterally, there is a broad band on each side of rich olive yellow, and the tips are grizzled yellow and black. Skull and teeth apparently not definitely distinguishable from those of *Tupaia ferruginea*. Zygomatic vacuity, large, widely open, about 5 by 2 millim. in the type.

Dimensions.—Head and body (c), 200 millim; tail (c), 140 (the *extreme* tip of the tail in the type is apparently wanting; the tail length in two other specimens is 127 and 153 respectively); hind foot, 41. Skull: basal length (c), 45·5; greatest breadth, 27; anterior rim of orbit to nasal tip, 22·5; interorbital breadth, 15; palate length, 27·8, breadth outside $\underline{\text{m}^2}$, 16·4, inside $\underline{\text{m}^2}$, 9·7; diastema between $\underline{\text{i}^2}$ and $\underline{\text{c}}$, 4·5, between $\underline{\text{c}}$ and $\underline{\text{p}^2}$, 1·1; front of i^1 to back of $\underline{\text{m}^3}$, 27.

Tupaia montana is most nearly allied to *Tupaia ferruginea*, Raff., and *Tupaia picta*, Thomas. It is distinguished from both by the colour of its tail, the former having this member dull annulated grey, and the latter brilliant rufous; it has also a shorter tail than either. From the former again it is separated by developing in old age a median dorsal black line, and from the latter, in which the line is present at all ages, by its less sharp definition, and also the greater uniformity of the general dorsal coloration.

Hab. Mount Dulit 5000-ft. (Type of species) (C. Hose). Mount Batu Song 3000-ft. (C. Hose).

TUPAIA MELANURA.

Tupaia melanura, Thomas, Ann. Mag., N. H. (6), ix., p. 251.

Size very small, less than in *Tupaia minor*. Fur very soft, close, and velvety. General colour dark olivaceous grey, finely sprinkled with yellow, a slight suffusion of dark rufous on the rump and base of tail. Face rather clearer olive; a short orange-coloured stripe above and below the eye, but not passing backwards towards the ear. No pale shoulder stripe present. Belly-hairs grey basally, washed terminally, from chin to anus, with bright orange. Outer sides of limbs like back, inner sides like belly; upper surfaces of hands and feet nearly black. Tail furred and coloured like body for about its basal inch and a half above and half inch below, but beyond that it differs from that of all other species by being quite cylindrical and short-haired, the hairs being closely adpressed and not forming a terminal pencil; in colour the short-haired part is deep jet-black throughout. Skull (Plate xix., figs. 4 and 5,) delicate, smooth, and evenly rounded. Zygomatic foramen reduced to a minute oval opening, which will scarcely admit the point of a needle. Palate without vacuities.

Dimensions (approximate, from skin)—Head and body, 125 millim; tail, 136; hind foot, 27·9; skull, basal length, 30; greatest length, 36; greatest breadth, 17·7; nasal length, 13; interorbital breadth, 10; intertemporal breadth, 14; palate length, 18·2; breadth outside $\underline{m2}$ 9·6; inside, $\underline{m2}$ 5·4.

Mr. Thomas writes in the P.Z.S., 1892, No. xvi., p. 225:—"This beautiful little species is the most interesting of the *Tupaiæ* obtained, as it forms a connecting link with the two species belonging to the genus *Dendrogale*. That genus was founded by Dr. Gray and recognised by Dr. Anderson in his recent review of the *Tupaiidæ*, mainly on account of its cylindrical tail, black and white banded cheeks, and the absence of the usual shoulder stripe. Now *Tupaia melanura* on the one hand has a tail even slenderer and more cylindrical than *Dendrogale murina* and *Dendrogale frenata*, and has no shoulder-stripe, while on the other its face-markings are quite as in *Tupaia*. One character, however, distinguishes *Dendrogale*, or at least *Dendrogale frenata*, from all the *Tupaia*, namely the extremely small size of the claws, both fore and hind; and so far as this character is concerned, *Tupaia melanura* is a true *Tupaia*, as it has claws quite as large in proportion as the other species. For the present therefore I consider it to be a *Tupaia*, and leave the validity of *Dendrogale* as an open question to be settled when further, and especially spirit, specimens are obtained."

The type of this pretty little tree shrew was obtained by me on the top of Mount Dulit at 5000-ft., living amongst the moss covered stunted jungle, and it is apparently a true mountain species, as I have since obtained other specimens, none of which were found below 3000-ft.

Hab. Mount Dulit (Type of species) 5000-ft. (C. Hose). Mount Batu Song 3000-ft. (C. Hose). Mount Kina Balu (A. Everett).

TUPAIA GRACILIS.

Tupaia gracilis, Thos. Ann. Mag. N. H. 1893.

Very similar in general appearance to *T. minor*, but much larger, and with longer hind feet.

Dimensions—head and body 160 mm., tail 170, hind-foot 38.

Hab. Sarawak (G. Doria). Apoh R. (A. Everett). Baram River (C. Hose).

GENUS DENDROGALE.

DENDROGALE MURINA.

Dendrogale murina, Müller and Schleg. p. 167. 1839-1844.

This tree shrew is easily distinguished from any of the *Tupaiæ* by the stripes on the face; the size is about that of *Tupaia melanura*.

Hab. Pontianak (Diard).

GENUS PTILOCERCUS.

PTILOCERCUS LOWII.

Ptilocercus lowii, Gray, P.Z.S. 1848, p. 23.

Blackish-brown, very minutely grizzled with the yellowish tips of the hairs: lips, lower part of cheeks, chin, and beneath yellowish: sides of the face inclosing the eyes black. Tail black; barbs white, except a few hairs near the scaly part, which are black.

Length $5\frac{1}{2}$ inches; tail, $6\frac{1}{2}$; hind-foot, 1. Skull: length, 1" 4"'.

Hab. Kuching (Type of species), (H. Low), Baram River (C. Hose), Lawas Mountains (A. Everett).

FAMILY ERINACEIDÆ.

GENUS GYMNURA.

GYMNURA RAFFLESI.

Gymnura rafflesi Horsfield and Vigor, Zool. Journ., iii., p. 248, pl. 8 (1827).

Tail about three fourths the length of the head and body, compressed towards the tip, naked scaly, the scales small and arranged in rings, between which short hairs project, becoming coarse bristles on the under surface, where the scales are more convex and more distinctly imbricate than above. Ears short, rounded, almost naked. Body clothed with hair of two kinds, the underfur soft and woolly, the longer hairs coarse bristles. Claws curved, not retractile. Skull very long and narrow, third upper premolar much larger than the second, and having three roots. The specimens from Sarawak differ from the typical form in being of white colour, only a part of the longest and strongest hairs on the trunk being black. The head, legs, and tail are pure white. This is merely a local variety. This white hedge hog is purely nocturnal, and lives under the roots of trees. Its food consist of insects. It has a very offensive smell, and is known to the natives as 'tikus bulan' or 'agi bulan.' *Tikus* in Malay means a rat, and *bulan* is the moon. Probably the name is given on account of its nocturnal habits. It is caught by the natives in spring traps called 'jirats,' but is not often eaten on account of its overpowering smell.

Hab. Baram River (C. Hose). Mount Dulit (C. Hose). Batu Song (C. Hose). Sarawak (A. Everett).

GENUS HYLOMYS.

HYLOMYS SUILLUS.

Hylomys suillus dorsalis, Müller and Schleg. Verhandl. Mamm. p. 153, pl. xxv., figs. 4-7, pl. xxvi., fig. 1 (1839-44);

Tail short, one sixth the length of the head and body, almost naked, and covered with small scales arranged in rings. Ears rather larger proportionately than in *Gymnura rafflesi*, rounded, almost naked. Body clothed with hair of three kinds, the inner fine, the longer coarse and bristly. Claws stout, but little curved.

Skull not so long as that of *Gymnura rafflesi*. Third upper premolar scarcely larger than the second. Colour above rusty brown, below pale yellowish white; the seminude portions of the limbs and tail brownish yellow. The hairs on the back are tipped with black. Dimensions—head and body 4·9 inches, tail 0·9, length of ear 0·6, breadth the same, hind foot 1. Length of skull 1·4, Zygomatic breadth 0·75.

In a paper on the mammals of Mount Kina Balu, by Mr. Thomas, on a collection made by Mr. John Whitehead, which was published in the P.Z.S., of April, 1889, page 229, I see he gives the name of *Hylomys suillus dorsalis* to a variety of this species found between the heights of 3000 and 8000-ft., the description of which is as follows. Essential character as in the typical variety, but with a more or less distinct black line running from between the eyes down the neck to the middle of the back.

Dimensions:—Head and body (c.) 116 millims; tail 16; hind foot 25.

Since all the five or six specimens of Hylomys obtained on Kina Balu show a black dorsal line, sometimes, it is true, faint and indistinct, but always present, Mr. Thomas believes this to be a valid geographical race, characteristic at least of Mount Kina Balu, if not of the whole of Borneo. It should be stated that Dr. F. A. Jentink, of the Leyden museum, where the type of *H. suillus* is preserved, believes it to be not worthy of separation from that animal.

The true *Hylomys suillus* has been recorded from Burma, the Malay Peninsula, Sumatra and Java.

FAMILY SORICIDÆ.

GENUS CROCIDURA.

CROCIDURA FULIGINOSA.

Crocidura fuliginosa, Blyth, J.A.S.B., xxiv., p. 362.

This is the common shrew of the low country, which is constantly found dead in the middle of the road. It ascends the mountains to the height of 2000-ft. Upper teeth 16. Size rather small. Muzzle thinly clad. Ears large, nearly naked. Feet thinly clad above. Tail long, three fourth the length of the head and body, or more, sub-cylindrical, very gently tapering (except in the rutting-season, when the basal portion is thickened), nearly naked, being thinly clad with very short hair, amongst which a few longer hairs are interspersed in the basal half only.

Colour—Bluey slate above and below. Tail thin with a few longer hairs at the base.

Hab. Kina Balu (J. Whitehead). (A. Everett)).

CROCIDURA MONTICOLA.

Crocidura monticola Pet.?

Hab. Sarawak (A. Everett).

CROCIDURA HOSEI.

Crocidura hosei, Thomas, Ann. and Mag. Nat. Hist., Ser. 6, vol. xi., 1893. See plate.

Size very small, scarcely exceeding that of the minute Indian species *C. perotteti*, *hodgsoni*, &c., and belonging evidently to the same group, which has not hitherto been recorded from the Malay region. Fur close, crisp, and velvety. Colour deep smoky grey, finely grizzled with whitish; belly rather paler; ears, feet, and tail dark brown. Tail very short and slender, closely haired, with a few longish hairs as usual scattered among the shorter ones.

Anterior part of first incisor short and thick; posterior cusp about equal in size to one of the two posterior unicuspids; first unicuspid large, its tip reaching the same level as the first incisor and the tip of the large premolar; second and third unicuspids about equal in size, the second slightly longer but slenderer than the third; fourth well developed, its tip just visible externally, and about equal in height to the basal process of the large premolar and about half the height of the third. Anterior lower incisor long and slender, without denticulations; second lower unicuspid just exceeding in height the anterior cusp of the large premolar.

Dimensions of the type (an adult skin):—

Head and body (probably stretched) 50 millim.; tail 20; hind foot 8·6.

Front of $\underline{\text{i. 1}}$ to back of $\underline{\text{m. 3}}$ 6·4; breadth of palate outside $\underline{\text{m. 1}}$ 4·1, inside $\underline{\text{m. 1}}$ 1·6; distance from tip of i. 1 to tip of the large premolar 3·1; length of lower tooth-row 5·8.

This species is larger than any of the Indian pigmy shrews, while it is far smaller than any other Oriental *Pachyura* as yet described. Its short tail seems also to distinguish it from all its allies.

I have only met with a single specimen of this tiny shrew.

Hab. Bakong River (Type of species) (C. Hose).

GENUS CHIMARROGALE.

CHIMARROGALE HIMALAICA.

Chimarrogale himalaica, Gray.

This is a velvety looking water shrew, and has its feet fringed with stiff bristles like the English water shrew, but it is about twice the size.

Hab. Kina Balu (J. Whitehead).

ORDER DERMOPTERA.

FAMILY GALEOPITHECIDÆ

GENUS GALEOPITHECUS.

GALEOPITHECUS VOLANS.

Lemur volans, L. Syst. Nat. i., p. 45 (1776).

Fur short, very fine and soft. Canines and the outer upper incisors and first lower premolar with low crowns. Anterior upper incisor trilobate (sometimes with four lobes).

Colour above varying from dark greyish brown to pale chestnut, alway overlaid, mottled, and blotched with silvery white. I have met with some specimens almost adult which were rufous all over except an occasional grey spot on the back.

Dimensions—Head and body, 16 inches, tail 9 inches.

They are usually found hanging to the trunk of a tree, the bark of which often resembles the back of the animal, and makes it very hard to distinguish.

The Dyak name is "Kubang plandok."

Hab. Baram River (C. Hose). Niah (C. Hose).

ORDER CHIROPTERA.

FAMILY PTEROPODIDÆ.

GENUS PTEROPUS.

PTEROPUS EDULIS.

Pteropus edulis, Geoffroy, Ann. Mus. xv., p. 90 (1810).

The largest bat known, the size being larger than that of *P. medius*. Ears naked, acutely pointed, longer than the distance from the eye to the end of the nose, narrower than those of *P. medius* (the breadth being only half the length), upper outer margin but slightly concave. The wing-membrane arises farther from the middle of the back, and the hairy back is much broader, otherwise the distribution of the fur is similar.

Colour generally similar to that of *P. medius* but rather darker. Head and breast rufous-brown, varying in tint ; hind neck and back between the shoulders paler yellowish or rufous-brown, or sometimes bright rufous ; back dark brown or black with a mixture of grey hairs ; lower parts either rufous-brown throughout, or the lower breast and abdomen nearly black with an intermixture of grey. Some specimens are black throughout.

Dimensions—Head and body 12 inches ; forearm over 8 inches. The expanse of wings is fully 5 feet.

The Dyaks call this species of bat 'Entamba' and the Malays 'Kaluong.'

It is usually seen in large numbers, feeding on the blossoms of the Mengwang and Durian trees.

Hab. Baram River (C. Hose). Niah River (C. Hose). Sarawak (A. Everett).

PTEROPUS HYPOMELANUS.

Pteropus hypomelanus, Temm.

GENUS CYNOPTERUS.

CYNOPTERUS SPADICEUS.

Cynopterus spadiceus, Thomas.

CYNOPTERUS LUCASI.

Cynopterus lucasi, Dobson.

CYNOPTERUS MARGINATUS.

Cynopterus marginatus, Geoffr.

CYNOPTERUS ECAUDATUS.

Cynopterus ecaudatus.

CYNOPTERUS MACULATUS.

Cynopterus maculatus, Thomas.

CARPONYCTERIS MINIMUS.

Macroglossus minimus, Geoffr.

EONYCTERIS SPELÆA.

Eonycteris spelæa, Dobson.

FAMILY RHINOLOPHIDÆ.

GENUS HIPPOSIDERUS.

HIPPOSIDERUS BICOLOR.

Hipposiderus bicolor, Temm.

HIPPOSIDERUS DORIÆ.

Hipposiderus doriæ, Peters.

HIPPOSIDERUS CERVINUS.

Hipposiderus cervinus, Gould.

HIPPOSIDERUS GALERITA.

Hipposiderus galerita, Cantor.

HIPPOSIDERUS DIADEMA.

Hipposiderus diadema, Geoffr.

GENUS RHINOLOPHUS.

RHINOLOPHUS MINOR.

Rhinolophus minor, Horsf.

RHINOLOPHUS AFFINIS.

Rhinolophus affinis, Horsf.

RHINOLOPHUS TRIFOLIATUS.

Rhinolophus trifoliatus, Temm.

RHINOLOPHUS LUCTUS.

Rhinolophus luctus, Temm.

FAMILY NYCTIRIDÆ.

GENUS MEGADERMA.

MEGADERMA SPASMA.

Megaderma spasma, Linn.

FAMILY VESPERTILIONIDÆ.

GENUS KERIVOULA.

KERIVOULA HARDWICKEI.

Kerivoula hardwickei, Horsf.

KERIVOULA PAPILLOSA.

Kerivoula papillosa, Temm.

GENUS VESPERTILIO.

VESPERTILIO MURICOLA.

Vespertilio muricola, Hodgson.

VESPERTILIO ADVERSUS.

Vespertilio adversus, Horsf.

GENUS HARPIOCEPHALUS.

HARPIOCEPHALUS SUILLUS.

Harpiocephalus suillus, Temm.

GENUS SCOTOPHILUS.

SCOTOPHILUS KUHLI.

Scotophilus kuhli, Leach.

GENUS VESPERUGO.

VESPERUGO DONRIÆ.

Vesperugo doriæ, Peters.

VESPERUGO TYLOPUS.

Vesperugo tylopus, Dobson.

VESPERUGO TENUIS.

Vesperugo tenuis, Peters.

VESPERUGO IMBRICATUS.

Vesperugo imbricatus, Horsf.

VESPERUGO STENOPTERUS.

Vesperugo stenopterus, Dobson.

FAMILY EMBALLONURIDÆ.

GENUS NYCTINOMUS.

NYCTINOMUS PLICATUS.

Nyctinomus plicatus, B. Hamiet.

GENUS CHIROMELES.

CHIROMELES TORQUATUS.

Chiromeles torquatus, Horsf.

GENUS TAPHOZOUS.

TAPHOZOUS AFFINIS.

Taphozous affinis, Dobson.

TAPHOZOUS LONGIMANUS.

Taphozous longimanus, Hardw.

TAPHOZOUS MELANOPOGON.

Taphozous melanopogon, Temm.

GENUS EMBALLONURA.

EMBALLONURA MONTICOLA.

Emballonura monticola, Temm.

EMBALLONURA SEMICAUDATA.

Emballonura semicaudata, Peale.

ORDER RODENTIA.

FAMILY SCIURIDÆ.

GENUS PTEROMYS.

PTEROMYS NITIDUS.

Pteromys nitidus, Desmarest. Nouv. Dict. d' Hist. Nat. vol. xxvii. 1818, p. 403 ;

Native name 'Kubang mirah.'

The colour of this flying squirrel may be described as a deep rich maroon-chesnut, inclining to black on the upper parts, the hairs being black-tipped on the back. On the base of the tail, which is bushy, the black tips are longer, and the chestnut portion of the hair becomes an obscure blackish chestnut, so that the tail, throughout by far the greater part of its length, is black, from the prevalence of the black tips and the dark underlying colour. The feet are concolorous with the body, which presents no trace of grizzling. The under parts are rich red-chestnut, liable however, to become paler, and the chin is blackish.

In the Leyden museum there is a large flying squirrel from Borneo, which appears to be a variety of this species. It is intermediate in its character between *Pteromys nitidus* and *Pteromys melanotis*, but unlike the latter, to the colour of which it has a general resemblance, it has no black points.

These flying squirrels usually frequent the clearings on which a few large trees are left standing. Living during the day time in holes about thirty to a hundred feet up the trunk, and coming out in the evening just at sundown. They breed in holes, not making a nest, and have from two to four young ones.

Hab. Baram river (C. Hose). Sarawak (C. Hose).

PTEROMYS PHŒOMELAS.

Pteromys phæomelas, Günther. Proc. Zool. Soc. Lond. 1873, page 413.

This is a brownish-black flying squirrel with a very glossy back. The hairs on the hinder half of the back have a narrow sub-terminal grey or white band which produces a very obscure, but minutely punctulated appearance to the fur. The upper surface of the parachute is less glossy, and the hairs have no pale bands. The under surface is sparsely clad with woolly chestnut-brown fur; the throat, the centre of the belly and the outer part of the membrane being most thickly clad. Cheek bristles are present, but they are feeble and few in number. The tail is bushy.

This dark coloured flying squirrel is rare. Its habits are like those of *Pteromys nitidus*. Native name 'Kubang arang.'

I obtained a very fine example of this species at Claudetown, which is now in the Sarawak museum.

Hab. Baram River (C. Hose).

GENUS SCIUROPTERUS.

SCIUROPTERUS PULVERULENTUS.

Sciuropterus pulverulentus Günther. Proc. Zool. Soc. Lond. 1873, p. 413, pl. xxxviii.

This is a small species, brownish black, with many of the hairs grizzled with yellowish, due to the presence of sub-apical yellowish bands, as the tips of the hairs are black. The basal two thirds of the fur are greyish, passing gradually into brownish black, which is succeeded by the yellowish ring, ending in the black tips. This grizzling extends on to the parachute, but not to the same degree as on the body and head. The grizzling of the feet is carried to such an extent that they are light brownish. The tail is distichous and bushy, the fur at its base being shorter than the remainder, and it is pale greyish-brown, the hairs being blackish at the tips.

Cheek-bristles are not observable. The ears are short and pointed, and covered behind with short black hairs. The incisors are pale yellow.

Length of body from muzzle to root of tail, 10 inches.
Length of tail - - - 9 inches.

This species is also rare. I have only obtained two specimens.

Hab. Claudetown (C. Hose) Mount Dulit (C. Hose).

SCIUROPTERUS HORSFIELDI.

Pteromys horsfieldi, Waterhouse. Proc. Zool. Soc. 1837, p. 87.

This species, which is a little larger than *S. genibarbis*, is recognised by the rich, uniform rufous-brown colour of the fur of the upper parts and tail, the latter being bright rusty beneath, bushy and distichous. The margin of the membrane and the sides of the face below the eye are reddish-yellow, and the dorsal surface of the parachute dark brown. On the upper surface of the body each hair is grey at the base; and the interspersed longer hairs, which are numerous, are bright brown or reddish-yellow at their apices. The fur is dense and woolly. On the under parts and inside of the limbs the hairs are yellowish-white and not grey at their bases. Cheek-bristles absent.

Body - - -	9.00 inches.
Tail - - -	8.25 „

I have only obtained one specimen of this small flying squirrel, which was shot near Claudetown. It is certainly very rare in the Baram district.

Hab. Baram river (E. Cox). Kuching (A. Everett).

SCIUROPTERUS SETOSUS.

Sciuropterus setosus, Temminck. Faun. Japan. Mamm. 1847, p. 49.

The type specimen, an adult female, was collected by Horner in the neighbourhood of Padang, in Sumatra. Dr. Anderson remarks in his work on the Zoology of Western Yunnan, page 294, "that he has examined the type of *Pteromys setosus*, which agrees with *Pteromys pearsonii* in the absence of cheek bristles and in the general characters, but the specimen is not fully grown, measuring only, along the back to the root of the tail, 4.75 and the tail 3.75. It is less rufescent than the adult, and the under parts are whiter, as are also the cheeks."

On the other hand, Dr. F. A. Jentink in the Notes from the Leyden Museum, Vol. xii. page 145, shows that they are different species and gives the measurements of the skulls of *Sciuropteruss etosus* and *Sciuropterus pearsonii* in millemetres, both being adult specimens.

	Sciuropterus pearsonii.	Sc. setosus.
Length of skull	41	30
Greatest breadth	27	17
Nasals	13.5	7
Palate	20	12
Diastema	9	6.5
Length of upper molar series	10	5.5

I have measured a spirit specimen in the Natural History Museum at South Kensington, which was collected in Sarawak by Mr. A. H. Everett. I also find a good skin in the Museum collected by me in 1880 from Claudetown. The measurements of Mr. Everett's specimen agree with those of Dr. Jentink. I believe this to be the smallest known species of Sciuropterus. The specimen I procured came out of a tree that had just been felled in the dense forest.

Hab. Baram River (C. Hose). Kuching (A. Everett).

SCIUROPTERUS GENIBARBIS.

Pteromys genibarbis, Horsfield. Zool. Researches in Java, 1824.

This is one of the smallest of the Southern Asiatic flying squirrels. The upper surface is pale yellowish-brown. The tail is markedly distichous, pale yellowish-grey at its base, the remainder being pale brown and the under surface somewhat rufous. The upper surface of the parachute is dark brown, the fore feet being pale yellowish brown, and the hind feet darker. The under parts are thickly clad with rather woolly hair, white, but with a faint yellowish tinge. The bases of the hairs on the sides of the belly and on the under surface of the parachute are slaty grey. The moustache is long and black. The sides of the face and neck are yellowish-white, with tufts of bristles on the cheeks.

Specimens of this flying squirrel from Borneo have been obtained on the Penrisen Hills, Sarawak, but as yet I have been unable to procure a specimen in Baram.

Hab. Sarawak (A. Everett).

GENUS SCIURUS.

SCIURUS EPHIPPIUM.

Sciurus ephippium, Müller Tijdschr. Over Nat. Gesch. 1838-39, p. 147.

This is the largest squirrel in Borneo, and it is common everywhere. It varies a good deal in colour at different times of the year, and has a tendency to become red as it gets to higher altitudes; for above 3,000-ft. a marked difference is noticeable the specimens.

Native name 'Enkrabak.'

Hab. Mount Dulit (C. Hose). Sarawak (A. Everett). Baram River (C. Hose).

SCIURUS HIPPURUS.

Sciurus hippurus, Is. Geoff. Étud. Zool. I. 1832, n. 6. tab. 6 ;

This well marked species approaches *S. erythraeus* in size. The lower parts and inside of the limbs are deep ruddy ferruginous; the head, sides of neck, shoulder, outside of fore limb, thigh, and outside of hind limb, being minutely speckled with white on a blackish ground ; feet black, the rest of the upper parts of the body with the base of the tail yellowish-rufous, punctulated with yellow and black ; tail bushy and black ; whiskers black. The specimens in the British Museum obtained by me from the Baram district are slightly different in the deepness of the colour, being a rich uniform chestnut on the back. This squirrel is usually seen on the ground, but takes refuge in the trees when frightened. It is common all through the low country in Sarawak, and does not ascend the mountains above 3000-ft.

Hab. Baram River (C. Hose). Mount Dulit (C. Hose). Sarawak (A. Everett).

SCIURUS PRYERI.

Sciurus pryeri, Thomas. Ann. Mag. Nat Hist. Sept. 1892.

This squirrel strongly resembles *Sciurus hippurus*, Geoff., in general appearance, although slightly smaller and more slenderly built, and agreeing precisely with that animal in the grizzled yellow colour of the back and the grey of the head and fore quarters, and their relative distributions on the anterior part of the body, but distinguished, firstly, by its wholly white instead of rich rufous belly ; secondly, by its hips being yellowish like the back, instead of grey like the head ; thirdly, by its feet being grizzled grey instead of black ; and finally, by its tail-hairs being broadly and conspicuously annulated with black and white, with white tips instead of being wholly black. Premolars $\frac{2}{1}$; incisors orange yellow, not darker above than below.

Dimensions of the type (an adult male in skin) :—Head and body 260 millim. ; tail 250 ; hind foot 54. Hab. [of the type (B. M. No. 92. 7. 19. 1).] Sapugaia River.

As yet this species has not been procured in the Baram district.

Hab. Sandakan (Pryer). Mount Kina Balu (J. Whitehead).

SCIURUS PREVOSTII.

Sciurus prevostii. Desmarest, Mamm. 1822, p. 335.

This is one of the most common squirrels in Borneo, and very changeable in colour, in the dry season it is quite black on the back with red on the belly. I have obtained several varieties of this pretty squirrel; the most usual is one with greyish black back, greyish white thighs, a broad yellowish white line from the axilla run along the side, and over the outside of the thigh to the heel; this tint is never really lost in any of the varieties, though in some cases it is very faint.

Length of body, 10·50; tail, not including hair, 10·30 inches.

This squirrel is found all through the low country, and on the mountains to 5,000-ft.

Hab. Baram River (C. Hose). Lawas (A. Everett). Mount Dulit (C. Hose).

SCIURUS NOTATUS.

Sciurus notatus, Boddaert, Elench. Animal, 1785, p. 119.

This common species, the Plantain Squirrel of Pennant, is represented on the mountains of Borneo by a blue-bellied type, without any trace of red or yellow on the undersides. The mountain specimens are smaller than those of the low country, and the tail has no trace of red hairs, but in every specimen that I have obtained of either sex, above 2,000-ft., the tail has bars across it.

In the low country this squirrel frequents the gardens and plantations and is particularly destructive in the cocoanut plantations, spoiling the young cocoanuts when they are about the size of a hen's egg.

The native name is "Tupai pinang."

Hab. Baram River (C. Hose). Sarawak (A. Everett). Mount Kina Balu (G. D. Haviland). Limbang River (C. Hose). Mount Dulit, 4,000-ft. (C. Hose).

SCIURUS EVERETTI.

Sciurus everetti Thomas. Ann. Mag. Nat. Hist. Aug. 1890.

Fur thick and soft, markedly more so than in the somewhat similar *S. tenuis*, Horsf., found in the same district. Colour uniform dark grizzled olive, rather darker than in *S. tenuis;* sides of cheeks, shoulders, and front of hips with a very faint fulvous suffusion. Under surface dirty greyish white, the hairs everywhere slaty grey for two thirds their length, then tipped on the throat and belly with dirty white, and on the chin and breast with dull fulvous. Ears short, rounded, not tufted or emphasized in colour. Tail unusually short, comparatively short-haired, almost cylindrical, the hairs ringed with dull fulvous and black. Skull small and lightly built, muzzle proportionally very long and narrow. Premolars $\frac{2}{1}$; molars small and delicate, their series on the two sides parallel, little bowed.

Measurements of the type (an adult skin) :—

Head and body 175 millim. ; tail, without hairs 109, with hairs 144 ; hind foot, without claws, 40 ; Skull, tip of nasals to bregma (centre of fronto-parietal suture), 36 ; zygomatic breadth, 24·5 ; interorbital breadth, 14 ; length of nasals, 15·7 ; breadth of nasals anteriorly, 5·2, posteriorly, 4 ; palate length, 24·2 ; diastema, 12 ; length of tooth series, 8·7.

A second specimen in the British Museum is rather larger, measuring :—Head and body, 180 millim. ; tail, without hairs, 118 ; hind foot, 40·5. This species is superficially by no means unlike *S. chinensis*, Gr., or *S. lokriah*, Hodgs., agreeing with both in its general size, and its uniform dull grizzled olive colour ; but it may be readily distinguished from either by its elongated muzzle, which allies it rather to the four following species. Of these, *S. laticaudatus* is separated by its larger size, shorter hair, browner colour, nearly white belly, and still longer muzzle ; *S. rufigenis* by the brilliant rufous of its cheeks and the underside of its tail ; *S. pernyi* by its similarly rufous tail ; and *S. berdmorei* by the black and white longitudinal stripes with which its body is ornamented. No other species that I can find has any close relationship to this form, which was discovered by Mr. Everett.

Hab. Baram River (C. Hose). Mount Penrisen, Sarawak, (A. Everett). Mount Dulit (C. Hose).

SCIURUS JENTINKI.

Sciurus jentinki Thomas. Ann. Mag. Nat. Hist. (5) xx. p. 129 (1887).

I have met with only two specimens of this squirrel on the mountains of the Baram district, and those only on Mount Dulit at 4000 feet.

Size about equal to that of *S. tenuis*, Horsf. General colour of upper surface yellowish grey, strongly suffused with orange on the head and along the centre of the back. Hairs dark slaty grey for four-fifths of their length, their tips yellow or orange. Face grey, but with white rim round each eye. Ears extremely short, rounded, their edges white or pale yellow, and standing out in a marked contrast against a patch of woolly black hairs situated just behind them on the sides of the neck. Hairs of chin, chest and belly slaty grey basally, dull yellowish white distally ; line of demarcation on sides quite gradual. Limbs coloured as in *S. tenuis ;* hind soles hairy for their proximal 8 millimetres. Tail slender, the hairs being comparatively short, only about 10 or 12 millim in length ; these hairs are broadly ringed with orange basally, and have a black subterminal and a white terminal band.

Incisors dark yellow above and below; premolars $\frac{2}{1}$; molars rather smaller and lighter than those of *S. tenuis*. Dimensions of specimen, a female, preserved in skin :—

Head and body 140 millim ; tail without hairs 103, with hairs 136 ; hind foot 32.5 ear, above crown 4·0.

Hab. Mount Kina Balu (J. Whitehead). Mount Dulit 4000-ft. (C. Hose).

SCIURUS BROOKEI.

Sciurus brookei, Thomas, P.Z.S. March, 1892, p. 225-226.

This pretty little squirrel is found on Mount Dulit between 2000 and 5000-ft. It is also found on Mount Batu Song and other mountains in the Baram district above the altitude of 2000-ft. The type was obtained by me in November, 1891, on Mount Dulit. It is about the size of *Sciurus lokriah*, Hodgs., or rather smaller; decidedly larger than *S. tenuis*, Horsf. General colour above plain olive grey, grizzled with yellow, but not so finely as in *S. tenuis*. Sides of body and outer and upper surfaces of limbs like the back, without the rufous suffusion characteristic of *S. tenuis*. Cheeks, anal region, and basal inch of tail below brilliant rufous. Chest and belly greyish white, the hairs grey basally, and dirty white terminally. Tail hairs broadly annulated with black and pale yellow.

Dimensions of type, an adult female in skin :—

Head and body 205 millim. ; tail, without hairs, 144 ; hind foot 37.

Skull: basal length (c) 37 ; bregma to nasal tip, 32 ; greatest breadth 25·6 ; nasals, length 13·2 combined breadth 7 ; interorbital breadth 15 ; diastema 10·6 ; palate, length 22, breadth outside $\underline{m1}$ 10·2, inside $\underline{m1}$ 6; front of $\underline{p4}$ to back of $\underline{m3}$ 7·4.

This squirrel belongs to a group of oriental species characterised by their dull grizzled olive-grey colour, unstriped sides, and annulated black and yellowish tails. For ornamentation some of the species have rufous patches on the head, shoulders, hips, or tail, but some are quite without them, and in all they vary much in their development.

Hab. Mount Dulit (C. Hose). (Type of species) Mount Batu Song (C. Hose).

SCIURUS LOWI.

Sciurus lowi, Thomas, Ann. Mag. N.H. (6) ix. p. 253, 1892.

Head and back a much darker shade than *Sciurus tenuis*, and with soft yellowish white fur on the belly, varying but little in colour from the chin to the root of tail ; tail much the same colour as the back, but with slightly more yellowish brown at the tips of the hairs. Size rather smaller than *Sciurus tenuis*. This is a low country species, only ascending the mountains to about 1000-ft.

Baram river (C. Hose). Mount Dulit 1000-ft (C. Hose). Mount Mulu (A. Everett).

SCIURUS TENUIS.

Sciurus tenuis, Blyth, Journ. As. Soc., vol. xvi., 1847, p. 874 (in part).

The head is concolorous with the back, sides of neck and thighs pale rufous not being much pronounced. Back olive brown, belly yellowish white. From tip of the muzzle to root of the tail measures 8 inches, and the tail 6, and with hair 8 inches. Common everywhere in Borneo, both in the low country and on the mountains to the height of 3000-ft.

Hab. Baram River (C. Hose). Penrisen Hills (A. Everett).

SCIURUS LATICAUDATUS.

Sciurus laticaudatus, Müller & Schlegel, Verhandl. 1839-44, pp. 100 and 215, figs 1, 3;

As remarked by Müller and Schlegal, its pelage has a strong resemblance to the pelage of *S. insignis*, having much the same character, except that it has no black bands. The coloration, as in *S. insignis*, is more murine than in any other Asiatic squirrel, except perhaps *S. davidianus*, and it is very variable in its intensity, varying from light to dark, almost blackish brown. It is about the size of *S. insignis* and the tail is shorter than the body, reaching to about the eye when laid forwards. The tail is moderately bushy, rather contracted at the base, but expanding towards the tip. The hairs are banded rather broadly with four alternate pale brown and dark brown bands, the last band being the darkest and the broadest, with a pale brown tip. The ears have the same form as in the squirrels, but the moustache is much more feeble. The under surface is nearly pure white in some, and rich orange-yellow in others.

This species, which is found in Borneo and the Malay Peninsula, presents a striking resemblance to a Tupaia. It is a very rare squirrel in Borneo, I have as yet met with but two specimens in the Baram district, both of which were found in the low country.

Hab. Baram River (C. Hose). Kina Balu (A. Everett).

SCIURUS HOSEI.

Sciurus hosei, Thomas Ann. Mag. Nat. Hist. for Sept. 1892. p. 215, 216.

This somewhat striking squirrel was found by me both on Mount Dulit and Mount Batu Song. It is undoubtedly a ground squirrel and only found on the mountains ranging between 2000 and 5000-ft. Its habits are similar to those of *Sciurus insignis* of the low country. I will here give Mr. Oldfield Thomas' description of the species.

"A striped squirrel of the size and somewhat the general appearance of *S. berdmorei*, Bly., but the muzzle short, as in the ordinary species. Ground colour of body olivaceous greenish grey, but this colour is only present in purity along the sides of the body and on the face, the nape of the shoulders being suffused with fulvous which narrows and brightens posteriorly into a defined dorsal fulvous line, on each side of which there are, firstly, a black, then a pale yellowish white, and then another black line. The resulting effect is not unlike some of the darker-coloured specimens of *S. tristriatus*, Waterh. (although with the centre line deep fulvous), or of some of the varieties of *S. berdmorei*. Under surface from chin to anus brilliant fulvous, the bases of the hairs whitish on the chest, greyish on the belly. Hands and feet grizzled with orange and black. Tail hairs broadly ringed with bright fulvous and black, the tips of the hairs fulvous. Premolars $\frac{2}{1}$, at least in the milk-dentition ; incisors deep orange-red above, rather pale below.

Dimensions of the type (a slightly immature male in skin) :—

Head and body, 245 millim ; tail imperfect ; hind foot, 42 ; combined length of three upper true molars 6·2 ; distance from front of $\underline{m1}$ to back of incisor 15·2.

Hab. Mount Dulit 4000-ft. (C. Hose). (Type of species) Mount Batu Song (C. Hose).

SCIURUS INSIGNIS.

Sciurus insignis, F. Cuv. Hist. Nat. des Mammif. Nov. 1821, pl. 223.

The young of this squirrel is dark brown, with a rather bright rufous tint on the front of the thighs, and, more or less also on their outsides, and on the shoulders. There are three narrow black streaks from the shoulder converging slightly towards the root of the tail. The under parts and the inner sides of the hind legs are dusky and tinted with the previous colour. The feet are dark, unpunctulated brown. The backs of the ears are clad with short hairs. The whiskers are black. The tail is dark brown but some of the hairs have very obscure greyish tips. The fur is very finely punctulated with rich yellow, the tips of the hairs being brown and their basal two-thirds greyish. The hairs of the tail, when pulled asunder, are also seen to be feebly annulated with yellowish brown and dark brown, the subterminal (basal) ring being of the former colour, and the free end of the hair very broadly blackish brown, occasionally with a greyish tip. In the adult, the sides of the animal are rufous-brown, lightest on the thighs. The area between the black dorsal lines, on the two anterior thirds of the trunk is much punctulated with yellowish, which also extends to the head. The hinder part of the back is rich red brown, the under parts are clear yellow white, the insides of the limbs are washed with rufous. Length from muzzle to tail 7·50 inches; tail reaching forwards to nearly the head. The skull has a rather narrow and pointed facial portion, and no great breadth between the orbits.

This is a true ground squirrel and is found all through the low country of the Baram district. It is fond of living on the slopes of hills, jumping about among the roots and leaves. I have frequently met with specimens which have lost part of their tails, and think that it is not unlikely that their tails are bitten off when fighting.

SCIURUS MELANOTIS.

Sciurus melanotis, Müller and Schlegel, Verhandl. 1839-44. p. 87.

The head is rather broad, thick, and blunt for the size of the animal, while the muzzle is short, and very broad at its origin. The ears are slightly pointed at their tips, and have a somewhat elongated form, being one-third longer than broad; they are clad with short yellowish hairs on the inside, and on the back with long black hairs which extend beyond their margins.

The colour is subject to considerable variation. In Java, the species is a dull pale yellow-brown, passing into bright olive, variegated with black, which is produced by the black annulation of the fur. A moderately broad white stripe runs along both sides of the head, below the eyes and ears, suddenly ceasing at the shoulders. It is bordered above to the ear, and along its lower margin till just below the ear, by a narrow black band. In some specimens, the whole of the hinder portions of the head and neck is occupied by a large spot, which is paler than the ground colour; but it is obsolete in other Sumatran examples, and those from Borneo are nearly alike, and differ only in the following details from the Javan animals. The colour of the upper parts is brighter, and merges, principally, on the head, into a nut brown. The under parts are pale yellow with a reddish tint, while in those from Java they are greyish-white passing into yellow. The hairs of the ears are longer than in the Javan animals, and the spot on the neck is more clearly defined, and is generally of a pale yellow tint.

The body measures - - -	3·50 inches.
Tail, without the hair - -	3·60 ,,

The specimen referred by Waterhouse to *S. soricinus* is in the British Museum, and it is a Javan specimen.

The little squirrel is found both in the low country in Borneo, and on the mountains to the height of 2000-ft. Usually two or three of them are seen together chasing one another round the trunks of the large forest trees.

Hab. Mount Dulit (C. Hose). Sarawak (A. Everett).

SCIURUS EXILIS.

Sciurus exilis. S. Müller.

The general colour of this tiny squirrel on the upper parts may be described as olive-brown, but the head and the back especially over the neck and shoulders, are more or less suffused with reddish, which, however, is not very prominent nor contrasting much with the general colour. The muzzle is yellowish, and there is a similar ring round the eye, but the sides of the face from the moustache backwards, and the sides of the neck resemble the sides of the body. The limbs also are concolorous with the body. The under parts are whitish or dusky, suffused, more or less, with rufous, and on the scrotum of the males with bright orange. The hairs of the tail have a broad basal band succeeded by a broad black band which is tipped with yellowish; the under surface of the tail being rather brightly worked with orange. The ears are of moderate size rounded, and clad with very short hairs. The whiskers are black. Nearly one half of the sole of the hind foot is clad.

The iris is brown; and the upper incisors are very pale yellow; the lower pair are nearly white. The skull is much arched behind; the facial portion is very broad at the base, becoming pointed towards the front, and moderately long.

This species is found in Borneo, Sumatra, and Malacca. Its native name is 'pukong'; it is common all through the low country, ascending the mountains to the height of 2000-ft. Like *S. soricinus* and *S. whiteheadi* it is usually seen running round the branches and trunks of trees making a curious little sqeaking noise with each movement, and whisking its tiny tail about with an air of importance.

Hab. Baram River (C. Hose). Sarawak (A. Everett). Mount Dulit (C. Hose). Mount Kina Balu (J. Whitehead).

SCIURUS WHITEHEADI.

Sciurus whiteheadi, Thomas Ann. Nat. Hist. Aug. 1887, p. 127; P.Z.S. 1889. p. 231, pl. xxiv.

This species is allied and very similar to *S. exilis*, Müll., but slightly larger, and with the ears, instead of being rounded and short haired, narrow, pointed, and with beautiful long black-and-white pencils of hair, nearly as long as the head, and standing out conspicuously from the general grey of the body. A white spot also present on the neck just behind the ear. Colour elsewhere precisely as *S. exilis*. Face without any trace of the black-and-white markings characteristic of *S. melanotis*, Müll. and Schl.

Skull very peculiarly shaped, with a short broad cranial and a disproportionally long and powerful facial portion, the distances from the tips of the nasals to a point between the anterior edges of the orbits 12·8 millim., as compared to about 10 millim. in *S. exilis*, and 11 millim. in *S. melanotis*, the latter an animal with the cranial part of the skull as large as, if not larger than, that of *S. whiteheadi*.

Dimensions of skin :—

Head and body, 90 millim ; tail, without hairs, 53 ; with hairs, 87 ; hind foot, without claws 24.5 ; ears, without hairs, 7 ; with hairs, 28.

This beautiful little squirrel is found on many of the mountains of Borneo between 2000 and 5000 ft. The type was discovered by Mr. Whitehead on Mount Kina Balu in 1887. I have since obtained specimens from Mount Dulit and Mount Batu Song. It frequents the dark forest, running rapidly round and round the branches of trees, with very mouse-like movements.

Hab. Mount Kina Balu (J. Whitehead). Mount Dulit (C. Hose). Mount Batu Song (C. Hose).

GENUS RHITHROSCIURUS.

RHITHROSCIURUS MACROTIS.

Sciurus macrotis, Gray, Proc. Zool. Soc., Lond. 1856, p. 341, pl. xlvi.

The ears are large, with a pencil of elongated hairs at their tips ; the feet are large and strong ; and the sides of the animal are laterally banded.

The general colour is dark chestnut-brown, very minutely punctulated ; the hind quarters, including the base of the tail, and the outsides of the fore and hind limbs, brighter ; the feet blackish. There is a brownish band from the axilla to the groin, with a yellowish-white band above it. The cheeks and inner sides of the limbs are pale brownish ; the chin, throat, and under parts, white. Tail broad and full, grizzled, and with long white tips to the hairs.

This handsome squirrel is rare, and as yet only found in Borneo. It is met with in the low country and ascends the mountains to about 2000 ft. It is constantly seen on the ground, and is the only Bornean squirrel which raises its tail over its back.

Native name, Krampu.

Hab. Baram River (C. Hose). Sarawak (A. Everett).

ORDER MURIDÆ.

GENUS CHIROPODOMYS.

CHIROPODOMYS PUSILLUS.

Chiropodomys pusillus, Thomas, Ann. and Mag. Nat. Hist. Ser. 6., vol xi. p. 345, 1893.

Size smaller than in *Ch. gliroides*. Ears and feet decidedly smaller and tail shorter than in that species. Fur crisp, close and velvety. General colour tawny fawn, head and centre of back darker, sides paler, outer sides of arms and legs like back, but the wrists and ankles greyish, a colour which also extends upon the metatarsus; fingers and toes white; under surface from chin to anus pure white; no darker markings on face; ears small, evenly oval, practically naked. Tail but little longer than the head and body combined, uniformly brown above and below, its terminal tuft of hairs of about the same thickness, but less extended and commencing more abruptly than in the allied species.

Skull smaller and rather more delicately built than in the other species and showing even more markedly the roundness, simulating immaturity, characteristic of the genus; supraorbital bead but slightly developed; anterior palatine foramina very short. Molars small, their structure as usual.

Measurements of the type (skin):—

Head and body 76 millim.; tail 81; hind foot 15·8; heel to front of last foot-pad 7·2; ear from notch 11·5.

Skull: upper length 22·2; breadth of brain-case 11·6; nasals, length 7·2; interorbital breadth 4·2; interparietal, length 4·2; breadth 9·2; anterior zygoma-root 2·1; diastema 6·2; anterior palatine foramina 2·7; combined lengths of $\underline{\text{m. 1}}$ and $\underline{\text{m. 2}}$ ($\underline{\text{m. 3}}$ is unfortunately lost) 2·5; length of lower molar series 3·1.

This species is founded on the specimen referred by Mr. Thomas in 1889* to *Ch. gliroides*, a reference mainly induced by the peculiar rounded and immature appearance of the skull; but this appearance has since proved to be a characteristic of the whole genus, and an examination of the teeth shows that the specimen is after all fairly adult. This being the case, the marked differences in the dimensions of the ears, feet, and tail will readily distinguish it from the older known species.

Hab. Mount Kina Balu (J. Whitehead). Mount Dulit (C. Hose).

CHIROPODOMYS MAJOR.

Chiropodomys major, Thomas, Ann. and Mag. Nat. Hist., Sec. 6, vol. xi., p. 334. 1893.

Colour and proportions very much as in *Ch. gliroides*, but

*P.Z.S. 1889, p. 235.

size conspicuously greater, especially so far as the skull is concerned (see dimensions below). Upper surface fawn, the bases of the hairs slate coloured; whole of lower surface pure white. Ears large, naked. Tail long, hairy, and pencilled as usual; uniformally brown above and below.

Skull with a flatter profile than in *Ch. gliroides*: anterior palatine foramina very short, ending half their own length in front of the molars. Molars broad and rounded.

Dimensions of the type (an adult female stuffed):—

Head and body 100 millim.; tail 109; hind foot 21·5; ear from notch 14.

Skull: upper length 30; breadth of brain-case 14; length of nasals 10; interorbital breadth 5·3; interparietal, length 5·2, breadth 10·4; anterior zygoma-root 3·1; palate, length 15·2; breadth outside m. 1 5·8, inside m. 1 3·4; diastema 8·5; anterior palatine foramina 3·8; length of upper molar series 4·4.

A second specimen from the same place agrees with the type in every respect.

Hab. Sadong, Sarawak (A. R. Wallace). Kina Balu (A. Everett).

GENUS MUS.

MUS RATTUS.

Mus rattus, Linn, Syst. Nat. i., p. 83 (1766).

Two specimens from Kina Balu mountain from an altitude of 8000-ft. have their fur long and soft, while those in those from 3000-ft it is short and harsh, so that it seems difficult to believe that both forms can be referable to the same species.

Dimensions—Head and body 5 to 8 inches; tail 5 to 9 or even more; hind foot without claws 1·2 to 1·5; ear 0·7 to 1.

Found everywhere both in the low country and on the mountains. Dyak name 'Chit,' Kayan name 'Lavo,' Malay name 'Tikus.'

Hab. Baram River (C. Hose). Mount Dulit 2000-ft. (C. Hose). Mount Batu Song 3000-ft. (C. Hose).

MUS ALTICOLA.

Mus alticola, Thomas, Ann. Mag. N.H. (6), ii., p. 408 (1888).

Fur mixed with flexible spines both above and below. General colour above a peculiar bluish grey, not speckled or grizzled, darker along the median line. Dorsal hairs and spines creamy white basally, gradually darkening to grey terminally. Underside pale yellowish white, the hairs and spines uniformly of this colour to their bases; the line of demarcation on the sides not very sharply defined. Hands and feet white, the hairs short and fine, fifth hind toe (without claw) reaching nearly to the end of the first phalanx of the fourth. Tail finely ringed, the rings averaging about 10 or 11 to the centimetre, short haired, sharply bicolor from base to tip, brown above, yellowish white below.

Dimensions—Head and body (probably stretched), 177 millim.; tail, 162; hind foot, 32; heel to front of last food-pad, 16; skull, tip of nasals to lambda (junction of sagital and lambdoid sutures), 34; nasals, length 15; interorbital breadth, 7·4; palate, length 19; length of anterior palatine foramina, 6; upper molar series, 5·8.

This species is most nearly allied to the Nepalese *Mus niviventer*, Hodgs., but may be distinguished by its unspeckled back, by the more gradual passage of the upper into the lower colour, and by its larger size.

Mr. Whitehead obtained the type on Mount Kina Balu at the height of 8600-ft.

MUS INFRALUTEUS.

Mus infraluteus, Thomas, Ann. Mag. N.H. (6), ii., p. 409 (1888).

Size large. Fur coarse and harsh, but not spinous. General colour dark greyish brown, the tips of the shorter hairs with a silvery lustre. The longer straighter hairs numerous, not markedly lengthened on the rump, uniformly black. Under surface a dirty yellowish brown, the tips of the straighter hairs dull orange, their base and the whole of the under-fur slaty grey. Ears small and rounded, naked. Hands and feet brown; last hind foot-pad elongate. Tail rather shorter than the head and body, thinly haired, dark brown above and below; rings of scales averaging about 8 or 9 to the centimetre. Skull stout and heavily built. Outer wall of infraorbital foramen evenly convex forwards. Palatine foramen about equal in length to the two anterior molars together, not reaching backwards to the front of m. 1. Teeth powerful; incisors broad; dark yellow in front above and below.

Dimensions—Head and body (c), 285 millim.; tail (extreme tip wanting), 235; hind foot, 51; heel to front of last foot-pad, 26; length of the same pad, 9·3; skull, tip of nasals to lambda, 51; nasals, length 21·8, breadth 6·5; combined breadth of upper incisors, 4·6; length of upper molar series, 10·7.

This fine rat has a certain similarity to the Indian Bandicoot rats (*Nesokia*), resembling them both in general external appearance and in the stout and heavy build of the skull and teeth. No species hitherto described can be mistaken for it, as all the Oriental rats which have external or cranial proportions at all similar are distinguished either by having elongated rump-bristles or parti-coloured white-tipped tails.

Hab. Mount Kina Balu 3000-ft. (type of species) (J. Whitehead).

MUS MARGARETTÆ.

Mus margarettæ, Thomas, Ann. Mag. N.H., Sec. 6, vol. xi., May, 1893 (see frontispiece).

Size, form, and general appearance very much as in *Vandeleuria oleracea*. Whiskers numerous and prominent, black. Ears small, oval, practically naked. Colour deep rufous chestnut, mixed on the back with the grey of the bases of the hairs, but clearing on the sides, where it seems to form a rufous lateral band. Chin, chest, and belly white. Hands and feet also white, but the metapodials with darker median patches. Thumbs prominent, opposable, with a large nail; claws of fingers short and curved. Hallux also opposable, its claw reduced to a minute conical point, not surpassing in length the pad below it; other toes all with their claws very short and curved, and surpassed in length by the prominent terminal pads. Soles naked, with six large rounded pads. Tail very long, slender, finely haired, almost naked; scales very small, averaging about seventeen to the centimetre, their colour a sort of pale greenish grey, the same above as below. Mammæ, 1—2=6. Palate-ridges, 3—5.

Skull with a very peculiar and note-worthy resemblance to that of *Chiropodomys*, agreeing with that of *Ch. gliroides* so closely that it is not until a close examination is made that the differences become apparent. General proportions short and broad, the brain-case especially broad and rounded.

Dimensions of the type obtained by Mr. A. Everett from the Penrisen Hills, Sarawak—Head and body, 76 millim.; tail, 144; hind foot, 19·7; ear, above head 11, from notch 13; length of head, 28.

This very remarkable species Mr. Thomas says will no doubt need in the future the erection of a special genus or sub-genus for its reception, and will perhaps prove to be congeneric with *Mus chiropus*, lately described by him in Ann. Mus. Genov. (2), x., p. 884 (1891), and p. 935 (1892), pl. xi., figs. 4—7, which also has the molar teeth of Mus combined with an opposable hallux and a general *Chiropodomys*—or *Vandeleuria*—like form. In his description of this pretty little tree mouse Mr. Thomas says, "I have taken the liberty of naming this beautiful little species, which looks as if it would make a most enchanting pet, in honour of Her Highness the Ranee of Sarawak, a lady whose interest in the zoology of that country is scarcely inferior to that of her husband the Rajah.

Hab. Penrisen Hills, Sarawak (A. Everett). Type of species.

MUS SABANUS.

Mus sabanus, Thomas, Ann. Mag. N.H. (5) xx. p. 269, 1887.

Fur short and fine, mixed with slender spines along the centre of the back. General colour rufous, mixed with brown along the top of the head and back, brighter and clearer on the cheeks and sides, the general tone very similar to that of *Mus jerdoni*. Whole of underside pure creamy white, sharply defined from the rufous of the sides. Outsides of limbs like sides, but rather greyer; inner sides, white; lower leg and ankles greyish brown all round. Hands and feet brown along the middle of their upper surfaces, their edges white, the contrast especially strongly marked on the feet, where a broad band of deep blackish brown passes along the centre, edged on each side with pure white.

Sole-pads large, smooth and prominent, the last one about three times as long as broad. Fifth hind toe, without claw, reaching to the end of the first phalanx of the fourth. Ears rounded, rather short, laid forward they barely reach to the posterior canthus of the eyes.

Tail enormously long, evenly finely haired, the scales, which are long, averaging from seven to nine to the centimetre, uniformly dark brown above and below throughout, but the hairs black for the proximal two thirds above only, elsewhere pure white. Mammæ, 2-2 = 8.

Dimensions of the type, which was discovered by Mr. J. Whitehead, an adult male, preserved as a skin:—Head and body, 280 millim.; tail, 340; hind foot, 43·5; ear, above head, 18; breadth, 18; heel to front of last foot-pad, 23; length of last foot-pad, 7·0. Skull: tip of nasals to centre of fronto-parietal suture ("bregma"), 36 millim.; nasals: length, 21; greatest breadth, 6·0; interorbital breadth, 7·7; length of upper molar series, 9·4.

One species, also a native of Borneo, has a certain superficial resemblance to the present one, although belonging to quite a different group of rats. This is *Mus muelleri*, Jent., of about the same size, and with a nearly equally long tail; but it may be readily distinguished by its coarse *Mus decumanus*-like fur, yellowish instead of rufous coloration, the less sharply defined white underside, and by the quite uniformly brown-haired feet and tail.

Mus sabanus is a hill species of rat, and fond of living in caves; some exceedingly fine specimens were obtained by Mr. Cox in the Niah Caves, in 1892, which apparently lived upon the food which was brought by people engaged in the collection of edible swallows' nests, which are abundant in many of the caves in Borneo. I also obtained two smaller specimens of the rat on Mount Dulit as high up as 5,000 feet. They were living in the moss which covers the top of that mountain, but they were evidently partial to rice, as one was shot in the act of making a hole in one of our rice bags.

Hab. Mount Kina Balu, 1,000-ft. (type of species) (J. Whitehead). Mount Dulit, 5,000-ft. (C. Hose). Niah Caves (E. Cox).

MUS EPHIPPIUM.

Mus ephippium, Jent.

Mr. Thomas, in his paper on the mammals of Mount Kina Balu collected by Mr. J. Whitehead in 1889, says, "It appears "rather doubtful whether this species is really distinct from "*Mus concolor*, Bly., found in Burma and in the Malay "Peninsula; but for the present I do not feel justified in "definitely uniting the two forms, and the Kina Balu in-"dividual clearly belongs rather to the Sumatran "*ephippium*" "than to its northern ally."

Hab. Mount Kina Balu (J. Whitehead).

MUS MUSSCHENBROECKI.

Mus musschenbroecki, Jent.

It is of considerable interest to find this species in Borneo, previously only known from Celebes, on a different side of the line, separating the Oriental from the Australian regions. Its occurrence here suggests that other members of the Oriental element in the peculiar Celebean fauna may also prove to have survived on the tops of the Bornean mountains. This rat was discovered for the first time in Borneo, by Mr. John Whitehead, on Mount Kina Balu, at the height of 3,000 feet.

Hab. Kina Balu (J. Whitehead). Mount Dulit, 2,000 feet (C. Hose). Penrisen Hills (A. Everett).

MUS HELLWALDI.

Mus hellwaldi, Jent.

This rat is spiny, with reddish yellow fur, black and white naked tail, at the end white all the way round. Feet very long and slender, $1\frac{1}{2}$ inches.

Hab. Batu Song (A. Everett). Dulit, 2,000-ft. (C. Hose). Penrisen Hills (A. Everett). Kina Balu (J. Whitehead). The type came from the Celebes.

MUS JERDONI.

Mus jerdoni, Blyth.

Very like *Mus hellwaldi*, but smaller. Foot about $1\frac{1}{8}$ inches; tail blackish at the end and rather hairy.

Hab. Mount Kina Balu (J. Whitehead).

MUS MULLERI.

Mus mulleri, Jent.

Large grizzled brown (not rufous) rat, with whitish yellow belly, uniform brown scaly tail. Foot nearly $1\frac{3}{4}$ inches.

Hab. Sadong (A. Everett). Dulit 2000-ft., Kina Balu (A. Everett.)

FAMILY HYSTRICIDÆ.

GENUS HYSTRIX.

HYSTRIX CRASSISPINIS.

Hystrix crassispinis, Günther. P.Z.S. 1876. p. 736, plate lxx.

This porcupine, which belongs to the same section as *H. javanica* and the allied species, is distinguished from all by the great size and length of the quills, all of which moreover are more or less distinctly grooved above, or at least provided with ridges. It is conspicuously smaller than *H. javanica*, but agrees with it in being covered everywhere with stiff spines, except on the foremost part of the head and abdomen. The largest quills are, in the middle, about twice as thick as an incisor. The prevailing hue of the head and fore part of the body is a greyish brown; but towards the large quills is white; the apical half black, with white tip. Slender quills white with black subcentral ring. Legs blackish.

This is the most common species of porcupine in Borneo and is known to the natives of Sarawak as 'Landak.' A curious soft concretion is found in the body of this animal and greatly prized by the Chinese as a medicine; the stone is called 'goliga landak.' The species lives in cavities under the roots of trees, and affords good sport when hunted with dogs, they can run at a great pace rattling the curious caudal quills as they rush through the forest. These animals, like the rhinoceros, feed upon the poisonous tuba root, which is almost certain death to any of the other animals in the Bornean jungle.

	in.	lin.
Length of body from tip of nose to root of tail -	17	0
Length of nose to ear - - - - -	3	6
Length of tail with terminal quills - -	6	0
Length of fore foot - - - - -	2	0
Length of hind foot - - - - -	3	0
Length of one of the largest quills - -	7	0
Length of one of the hollow caudal quills -	6	0 "

This porcupine is found on the mountains up to 2000 ft., and is common in the low country.

Hab. Baram River (C. Hose). Sarawak (A. Everett). Rijang River (H. B. Low). Mount Dulit (C. Hose).

HYSTRIX MÜLLERI.

Hystrix mülleri, Jent., notes from the Leyden museum.

This porcupine is like *Hystrix crassispinis*, but distinguished from it by its black belly and somewhat different caudal quills. The skulls also differ, but the size of the animals are much the same.

This porcupine is found both in the low country and on the mountains to the height of 2000-ft. The Dyak name is 'Landak Dudul.' Kayan name, 'Kalong.'

Hab. Mount Dulit, (C. Hose). North Borneo (A. Everett).

GENUS TRICHYS.

TRICHYS LIPURA.

Trichys lipura, Günther, Thomas. Proc. Zool. Soc. Lond. 1889, p. 75. *Trichys guentheri*, Thomas, P.Z.S. 1889, p. 235.

All the upper and lateral parts of the body are densely covered with flat flexible bristles of moderate length, grooved on the upper as well as the lower surface. Underfur very scantily represented by fine woolly hairs; and on the rump some long hair-like bristles project beyond the flat ones. On the head and abdomen the bristles are replaced by flat stiff hairs. In the external form and structure of the head, ears, and feet there is no marked difference from *Atherura*. The general tint of the upper parts of the animal is brown, each spine being white at the base, and brown towards the point. On the sides the brown colour gradually passes into the white of the lower parts. Hairs of the long moustache black, the longest hairs having whitish terminations.

	in.	lin.
Total length of body without tail -	17	0
Distance from nose to ear - -	3	0
Length of ear - - -	1	0
Length of sole of fore foot - -	1	6
Length of sole of hind foot - -	2	6
Length of one of the longest flat spines	1	8

The tail of a full grown specimen of which the skin measured 15 inches without the tail, is 8½ inches, slender, longer than one half of the body and head, covered with spines for about one inch of its basal portion. Nearly in the whole of its length it is covered with rhombic scales of relatively large size, and arranged regularly in oblique series or rings. A short fine hair, which is never spinous as in *Atherura macrura*, starts from the base of each scale and lies closely adpressed to its median line, giving to the scale the appearance of being keeled (like the scale of a snake). Towards the end of the tail the hairs become longer, and the terminal quills are much elongated, 2-3 inches long, and compressed with a shallow groove, like blades of grass, only much narrower, and form a thin bundle. The majority are truncate at their extremity and hollow. These quills, therefore, differ much in shape from those of *Atherura*, and are in fact a less developed form of the caudal quills of other porcupines. This species is unable to produce the rattling or quivering noise which the more highly specialized forms of porcupine make under the influence of fear or anger.

Dr. Günther described a tail-less specimen of this porcupine as *Trichys lipura*, but on the discovery that it possessed a tail Mr. Thomas re-named it *T. guentheri*. I had the good fortune to procure a mother and young one which put an end to any more doubt as regards the normal possession of a tail by this porcupine, the young one having a tail and the mother no sign of one. The specimens are now in the Natural History Museum at South Kensington. The natives of Sarawak call the porcupine "ankis," and say that there are two species, but the reason is simply because they see some with a tail and some without. However I should not be surprised to find that *Atherura* also occurs in Sarawak. Kayan name, 'Buka.'

Hab. Baram River (C. Hose). Sarawak (A. Everett). Rijang River (H. B. Low). Mount Dulit (C. Hose). Mount Batu Song (C. Hose).

ORDER UNGULATA.

FAMILY ELEPHANTIDÆ.

GENUS ELEPHAS.

ELEPHAS INDICUS.

Elephas indicus, Cuv. Règne An. I. p. 231 (1817).

Skin nearly naked. Tail with a row of long coarse hairs for a few inches before and behind and round the end only. Five hoofs normally on each fore foot; four hoofs on each hind foot. The number of ridges in each molar, from the first to the last, is 4, 8, 12, 12, 16 and 24, with slight variation. Colour, blackish grey throughout.

Dimensions.—Height at the shoulder in adult elephants is almost exactly twice the circumference of the fore foot. Adult males do not, as a rule, exceed 9 feet, females 8 feet, in height.

The elephants in Borneo are only met with in the northern portion of the island. They are very destructive to the gardens. They are called by the Malays 'Gajah.'

Some fine specimens of skulls of these animals are preserved in the Sarawak Museum.

FAMILY RHINOCEROTIDÆ.

GENUS RHINOCEROS.

RHINOCEROS SUMATRANUS.

Rhinoceros sumatranus, Raffles, Tr L. S. xiii, p 268 (1820).

This is the smallest of living rhinoceroses and the most hairy, the greater part of the body being thinly clad with hair of some length, and there being hair of considerable though varying length on the ears and tail. The two horns are some distance apart at the base; both are slender above, and the anterior horn, in fine specimens, is elongate and curved backwards. The skin is usually rough and grandular; the folds, though much less marked than in the one-horned species, are still existent, but only that behind the shoulders is continued across the back. Incisors in adults $\frac{2}{2}$, the lower pair lateral, large, and pointed; sometimes lost in old animals.

Colour varying from earthy-brown to almost black; hair of body brown or black.

Dimensions—Somewhat variable. The type of *R. lasiotis* was 4-ft. 4-in. high at the shoulder, and 8 feet long from snout to root of tail; its weight about 2000 lbs. (*Anderson*). An old female from Malacca was only 3-ft. 8-in. high. The average height of adults is probably 4 feet to 4 feet 6-in. The largest known specimen of the anterior horn measures 32 inches over the curve. Skull 20 inches in basal length, 11·25 in zygomatic breadth.

Varieties—Specimens from Chittagong and Malacca were living at the same time in the Zoological Society's Gardens, London, in 1872; and the former was distinguished by Sclater as *R. lasiotis* on account of its larger size, paler and browner colour, smoother skin, longer, finer, and more rufescent hair, shorter and more tufted tail, by the ears having a fringe of long hair but being naked inside, and above all by the much greater breadth of the head. Unquestionably the differences are considerable; but by far the most remarkable—the shape of the head—was shown by Blyth to be variable in both *R. unicornis* and *R. sondaicus*, for he figured and described a broad and a narrow type of each (J. A. S. B. xxxi, p. 156, pls. i–iv) as well as of *R. Sumatrensis*. The other distinctions scarcely appear to me of specific value, and I am inclined to regard the two forms as varieties only.

Habits—Very similar to those of the other species; this rhinoceros inhabits forests and ascends hills to a considerable elevation, having been observed 4000 feet above the sea in Tenasserim by Tickell. It is a shy and timid animal, but easily tamed even when adult. The horn is valued by the Chinese for the purpose of medicine; and it is occasionally met with in the interior of Borneo, but it is rare in the low country.

FAMILY BOVIDÆ.

BOS SONDAICUS.

Bos sondaicus, Müller & Schleg. Verhandl. p. 197, pls. xxxv–xxxix (1892).

This animal appears to be slighter than the gaur, with the legs longer in proportion and the dorsal ridge less developed. The tail descends below the hocks. The dewlap is of moderate size. The head is much more elongate, the forehead not concave, the horns smaller, cylindrical in the young, flattened towards the base in adults, and curving outwards and upwards at first, and towards the tips somewhat backwards and inwards.

Colour—Cows and young bulls have the head, body, and upper portions of the limbs bright reddish brown, approaching chestnut, old bulls are black ; in both sexes the legs from above the knees and hocks, a large oval area on the buttocks, extending to the base of the tail but not including it, a stripe on the inside of each limb, the lips, and the inside of the ears are white. Calves have the outside of the limbs chestnut throughout and a dark line down the back.

Dimensions—According to S. Müller, a full-grown Javan bull measured 5-ft. 9½-in. high at the shoulder, the length of the head and body was 8-ft. 6-in., and of the tail 3-ft. The largest Burmese specimen recorded was 16 hands high (5-ft. 4-in.). A skull from Java in the Indian Museum, Calcutta, has horns measuring 30 inches long by 17 inches in circumference at the base. This is unusually large. A male skull from Borneo in the British Museum measures 17·75 inches in basal length by 8·75 in zygomatic breadth.

So far as is known its habits are similar to those of *Bos gaurus*, except that *B. sondaicus*, from the greater proportional length of the legs, is probably less of a climber and more restricted to the plains of high grass.

Hab. Baram River (C. Hose). Niah River (C. Hose).

BOS BUBALUS.

Bos bubalus, L. Syst. Nat. i. p. 99 (1766) ; W. Sclater, Cat. p. 129.

General form heavy, body massive, legs thick and short, hoofs large. Tail reaching the hocks (but, I think, variable in length). Ribs 13 pairs. Hair on the body very thin, especially in old animals. Muzzle large and square. Head carried very low.

Skull elongate, nasals long, forehead nearly flat. Horns very large, flattened, transversely rugose, trigonal in section, tapering slowly and gradually from the base, curving at first upward, outward, and slightly backward from the plane of the face, the curve increasing towards the ends, where the horns curve inwards and a little forwards. The horns depart but little from one plane throughout. In some (*macrocerus* of Hodgson) the horns are almost straight till near the end, where they turn more rapidly upward.

Colour throughout dark ashy, almost black. The legs are sometimes whitish; in some tame forms the legs are white to the same height as in the Gaur. Horns black.

Dimensions—According to Jerdon (who probably took the figures from Hodgson) and others, the wild buffalo measures in height up to 6½ feet, and in length from snout to root of tail 10½. Kinloch, however ('Large Game Shooting,' ed. 2, pp. 88, 91), doubts if any exceed 5-ft. 4-in. in height (16 hands), and gives the following measurements of a good-sized bull: height 5-ft., length from nose to root of tail 9-ft. 7-in.; tail 3-ft. 11-in.; girth 8-ft. 3-in.; length of horns from tip to tip round curve 8-ft. 3-in. This is a common way of measuring buffalo horns. The longest recorded horns known, a pair in the British Museum, measure 78½ inches each, which would give an outside sweep of about 14 feet. Cows' horns are longer than bulls', but of less girth.

Habits—The wild buffalo keeps chiefly to level ground and is generally found about swamps. It haunts the densest and highest grass-jungle or reeds, but is also found at times in open plains of short grass, or amongst low bushes, but very rarely in tree-forest. Buffaloes associate in herds, often of large size. I have seen 50 together, and have heard of much larger assemblages. They feed chiefly on grass, in the evening, at night, and in the morning (probably morning and evening as a rule), and lie down, generally in high grass, not unfrequently in a marsh, during the day; they are by no means shy, nor do they appear to shun the neighbourhood of man, and they commit great havoc amongst growing crops. Whether wild or tame they delight in water, and often during the heat of the day lie down in shallow places with only parts of their heads above the surface.

Few, if any, tame animals have changed less in captivity than buffaloes. Unlike the yak and gayal, they never breed with tame cattle *(B. indicus)*, although the cows often pair with wild bulls of their own species. Tame buffaloes are chiefly kept for milk and for draught. They have been introduced throughout many of the warmer parts of the Old World, and even in Italy, whither they were brought in the sixth century (Griffith's Cuvier, iv, p. 381). Both wild and tame rut in autumn; the females gestate for 10 months (10 months and 10 days according to some) and bear one or two young in summer. Native name 'Krebau.'

Hab. Miri River (C. Hose). Baram mouth (E. Cox).

FAMILY CERVIDÆ.

GENUS CERVULUS.

CERVULUS MUNTJAC.

Cervulus muntjac, Brooke, P.Z.S. 1874, p. 38, 1878, p. 899.

Colour deep chestnut, becoming darker on the back and paler and duller below. Face and limbs brownish, a black line along the inside of each horn-pedicel and for some distance inside the facial rib; this line in the female ends above in a slight tuft. Chin and upper throat, lower abdomen, lower surface of tail and inside of thighs white; a whitish mark in front of the digits on each foot. Axils whitish. A dark brown variety has been found near Darjiling by Kinloch, and a still darker form is figured in Hodgson's MS. drawings. Young spotted.

Dimensions—Height at shoulder 20 to 22 inches; length of head and body about 35; tail, with hair, 7. A male skull measures 7 inches in basal length and 2·7 in breadth across the orbits. The horns from the burr (pedicel not included) rarely exceed 5 inches in length, and are generally 2 or 3 inches, on pedicels 3 to 4 long, but horns of 11 inches are said to have been measured. Weight of a male 38 lb.

Habits—The rib-faced deer is a solitary animal, usually found singly or in pairs. It keeps in thick jungle, only leaving the forest to graze on the skirts of the woods or in abandoned clearings. It has a wonderful way of getting through the thickest underwood, and it runs in a peculiar manner with its head low and its hind quarters high; when not alarmed, as Colonel Hamilton observes, it steps "daintily and warily, lifting each leg well above the grass or leaves."

The call of this species, from which the common name of "barking deer" is derived, is at a little distance very like a single bark from a dog, and is very loud for the size of the animal. It is often repeated at intervals, usually in the morning and evening, sometimes after dark, and I have heard it in Borneo very late in the morning and again in the afternoon, in the cold weather, which is the rutting-season. It is uttered by the animal when alarmed, as well as when calling its mate.

I obtained a very dark coloured specimen of this deer on Mount Dulit, at 3,000 feet. Dyak name, 'Kijang.'

Hab. Baram River (C. Hose). Niah (C. Hose). Mount Dulit, 3,000-ft.

GENUS CERVUS.

CERVUS EQUINUS.

Cervus equinus, Cuv. *Ossemens Fossiles*, p. 45, pl. v, figs. 37, 38, 46 (1823).

The largest Indian deer. Ears large. Hair coarse. Neck and throat of the adult male covered with long hair forming an erectile mane. Muffle large. Orifices of infraorbital glands very large and capable of eversion. Tail moderate. Interdigital glands wanting, according to Hodgson. Molars markedly hypsodont, with small accessory columns. A deep lachrymal fossa; auditory bulla slightly inflated and rugose. Horns each normally with but three tines and very rarely bearing more, irregular points or sports being less common than in most deer; the brow-antler meets the beam at an acute angle; the two upper tines generally sub-equal in Indian heads, but very variable in form and proportion.

Colour almost uniform dark brown throughout, sometimes greyer, sometimes with a slight yellowish tinge, scarcely paler below. Females and young paler and more rufous than males. Chin, inside of the limbs near the body, lower surface of the tail, and inner parts of the buttocks yellower, sometimes dull yellowish white. Young not spotted at any stage. Some old males are very dark-coloured, almost black or dark slaty grey.

Dimensions.—Height at shoulder of males 48 to 56 inches, and it is said even more; length 6 to 7 feet, tail 12 to 13 inches, ears 7 to 8. Females are smaller. A male skull measures in basal length 14·2 inches, extreme length 15·7, orbital breadth 6·7.

Habits.—Although it does not shun the neighbourhood of man to the same degree as *Bos sondaicus* does, it is only common in wild tracts of country. It comes out on the grass slopes, where such exist, to graze, but always takes refuge in the woods. It is but rarely found associating in any numbers; both stags and hinds are often found singly, but small herds from four or five to a dozen in number are commonly met with. Its habits are nocturnal; it may be seen feeding in the morning and evening, but it grazes chiefly at night, and at that time often visits small patches of cultivation in the half-cleared tracts, returning for the day to wilder parts, and often ascending hills to make a lair in grass amongst trees, where it generally selects a spot well shaded from the sun's rays. It feeds on grass, especially the green grass near water, and various wild fruits, of which it is very fond, but it also browses greatly on shoots and leaves of trees. It drinks, I believe, daily, though Sterndale doubts this; it certainly travels long distances to its drinking-places at times. The rutting-season is about October and November.

Dyak name 'Rusa.' Kayan name 'Payoh.'

Hab. Baram River (C. Hose). Niah (C. Hose). Mount Dulit, 2,000-ft. (C. Hose). Ridan River (E. Cox).

FAMILY TRAGULIDÆ.

GENUS TRAGULUS.

TRAGULUS NAPU.

Tragulus napu, A. Milne-Edw. An. Sci. Nat. (5) ii, p.p. 106, 158, pl. ii, fig. 2, pl. viii.

A naked tract on the throat, the tarsus naked behind, and the tail long as in *T. javanicus*. Size larger.

Colour—Upper parts yellowish or rufous-brown, sides greyer. Hair on back light brownish orange with black tips, no subterminal pale ring. On the sides the basal portion of the hair is whitish. Forehead and nape blacker, but the borders of the black area ill-defined. Lower parts white, generally a brown median line on the breast, the chest and lower abdomen white and an intermediate tract brownish. Throat and fore neck brown, with 5 white bands more or less distinct, a median band on the chest and two oblique lines on each side in front on the throat. The white lines often become blended together. Rump rufous: tail brown above, white below.

Dimensions.—Height 13 inches, nose to root of tail 28, tarsus and hind foot 5·6 to 6, tail 5. I have been unable to obtain the measurements of an adult skull; those of the figure in Milne-Edwards's paper are:—extreme length 4·5 inches, basal length 4, breadth 1·9, but these are probably small.

Dyak name, 'Plandok,' Kayan name 'Planok.'

This little deer has been known to kill chickens in the poultry yard, and carry them away into the jungle.

Hab. Baram River (C. Hose). Niah River (C. Cox). Mount Dulit, 2000-ft. (C. Hose).

TRAGULUS JAVANICUS.

Tragulus javanicus, A. Milne-Edw. t. c. pp. 103, 157, pl. ii, fig. 1.

A naked glandular area beneath the chin, between the rami of the mandible; tarsus naked behind throughout, carpus almost naked behind. Tail long.

Colour—Above brown, more or less rufous. Back in old individuals nearly black, but always more or less mixed with rufous or yellow, from some of the hairs having a yellow ring near the end. Hair at base light brown. Sides paler ; nape and upper surface of neck almost or quite black, contrasting with the light brown of the sides. Lower parts white, variously mixed with light rufous and usually with a median narrow brown or rufous line throughout the breast, in front of this is a brown cross band and on the fore neck an arrowhead-like brown mark, sometimes imperfect, with three white stripes, one median, within the arrow-head, the other two diverging, one on each side, outside of it ; the last two joining on the throat. Rump rufous, inside of thighs and intermediate space always white ; tail rufous-brown above, white below.

Dimensions—The largest adults measure : nose to root of tail 18·5 inches, tail 3 *(Cantor)*, tarsus and hind foot 4·4 to 5. Basal length of a male skull 3.4, extreme length 3·05 ; zygomatic breadth 1·0.

Dyak name ' Kamaya panas.' Kayan name ' Planok.'

Hab. Baram River (C. Hose). Suai River (C. Cox). Mount Dulit, 2,000-ft. (C. Hose).

FAMILY SUIDÆ.

GENUS SUS.

SUS BARBATUS.

Sus barbatus.

The pig is common everywhere is Borneo, and very destructive to the gardens, whenever there is plenty of fruit about, this animal comes in large numbers. They, however, afford very good sport, as will be seen in my short account of a days pig hunting taken from the *Field* of April, 1893 :—

"WITH SPEAR AND HOUND IN BARAM, SARAWAK.

It is six o'clock, on a bright spring morning ; all the men are up and hastily snatching a few mouthfuls of food, and vainly endeavouring to drink the hot coffee, which endeavour results in curses and scalded mouths. The hounds have had a small meal an hour before, and the Dyak hunters, spear in hand, are anxiously waiting for the signal to start. There is something refreshingly pre-historic about the sport in which we purpose indulging to-day. We discard the appliances of civilization as represented by the rifle and the 12-bore breechloader, and set out for the chase with much the same weapons, and in much the same manner as did our ancestors in the neolithic age. In our method and equipment we furnish an argument against those who hold that in our sports we take care to have overwhelming odds against the lower creation; for surely even they would not maintain that spear and hound against the wild boar's tusks give an undue superiority, or that to a man on foot in the jungle they are aught else than 'the weak one's advantage fair.' 'Now then you fellows,' cries the master of the hounds (Mr. E. A. W. Cox) from below, 'we must be off if we intend to make a day of it!'

The mist is gradually rising from the river as the warm sunbeams lighten the atmosphere. The birds are whistling all around, and the pretty bubbling noise of the Wa-Wa monkey harmonises with the morning symphony of nature newly awakened from sleep. We follow a jungle path for about half a mile, when suddenly the welcome music of the pack tells us that something has been disturbed. It is easy to distinguish by the barking of the dogs what game is afoot, as their bark when after a deer is quite distinct from that when after a pig, and these sounds are again different from their cry when in chase of other animals. For one moment we listen for the direction; then everyone bounds off through the jungle, tripping over roots and logs and brushing past thorns, quite unconscious of bruises and scratches in the wild excitement of the chase. Presently the barking changes, denoting to the practised ear that the pig is brought to bay. Now comes a race for first spear, and each, wild with excitement, strives his utmost to obtain the honour. If the pig is an old one he fights very hard, and often doubles up the spear in his huge jaws. The dogs soon become very artful in dealing with the quarry, and, in consequence, are seldom killed. Hanging up and removing the inside of the boar is the work of a few minutes, and then another start is made. Passing over undulating ground, having hardly recovered yet our breath, the cries of 'Rusa!' (deer) burst forth from the excited Dyaks. Away we go again, down the ravine and over the rocks, slipping and hurting our shins on the huge boulders; on up the little hills, and down again through a valley ending in a swamp, and here at last we find the deer, which, being surrounded by dogs, and unable to move in the marsh, is quickly despatched.

We throw ourselves down for a rest whilst the natives quarter the stag. Each man then hoists on his back as much as he can manage to carry, and we retrace our steps to the beaten track. Two more pigs afford us the pleasure of a good run before we are willing to return. Men and dogs are thoroughly tired; but the Dyaks, struggling as they are under the weight of their heavy load, yet lament the escape of one monster pig, of which the dogs lost the scent. Greatly fatigued and pleased with our sport, we return, and, after a refreshing bath and a good meal, we throw ourselves back in our lounge chairs, and hold a long, lingering discussion on the day's proceedings and on the delights of the chase."

The Dyak name is 'Janni,' Kayan name 'Baboi.'

Hab. Baram River (E. Cox). Mount Dulit 4000-ft. (C. Hose).

ORDER SIRENIA.

FAMILY MANATIDÆ.

GENUS HALICORE.

HALICORE DUGONG.

Halicore dugong, Illiger, Prod. p. 140; Gray & Hardw. Ill. Ind. Zool. ii, pl. xxiii; Blyth, J. A. S. B. xxviii, pp. 271, 483, 494; id. Cat. p. 143; id. Mam. Birds Burma, p. 53; Jerdon, Mam. p. 311; W. Sclater, Cat. p. 326.

Dimensions—Extreme length of adults 8 to 9 feet, usually 5 to 7; much larger dimensions are given in books, but are open to doubt. In a large specimen 8-ft. 6-in. long and 6-ft. in circumference, the pectoral fins were 16 inches long and 8 inches broad, and the breadth of the tail from tip to tip 31. The skull of a male from Ceylon measures 14.5 inches in basal length and 8.5 in breadth.

Distribution. The shores of the Indian Ocean from E. Africa to Australia for about 15 degrees on each side of the Equator. Dugongs have been observed on the coast of Malabar, the north-west coast of Ceylon from the Gulf of Calpentyn to Adam's Bridge, around the Andaman Islands, and in the Mergui Archipelago,

Habits—Formerly dugongs were said to be found in large herds of some hundreds of individuals, and to be in places so tame as to allow themselves to be handled. As their flesh is by all accounts excellent and their fat yields a clear limpid oil of great value, they have everywhere been hunted and are now in most places rare. They are said to be slow in their movements and unintelligent. Their food consists of marine algæ. They haunt shallow bays, salt-water inlets, and mouths of estuaries, but do not, like the Manati, ascend rivers. The female gives birth to but one young at a time, and is said to hold it with her pectoral fin. Some writers have suggested that the dugong has given rise to the myth of the mermaid (hence, indeed, the name *Halicore*); but it should be remembered that stories of beings half man or woman, half fish, are as common in temperate as in tropical seas, and that some of them are more ancient than any European knowledge of the dugong.

Hab. Bornean coast.

ORDER EDENTATA.

FAMILY MANIDÆ.

GENUS MANIS.

MANIS JAVANICA.

Manis javanica, Desmarest, Mamm. p. 377 (1822).

Form more slender than in either of the proceeding species and tail generally longer. Fore claws but little longer than the hind, never more than half as long again. Scales longer, more pointed behind, and rather less closely adpressed, the posterior edges chipped, not worn, and with a median keel frequently visible in adults, especially on the tail, sides, and limbs; 15 to 19 rows (usually 17) round the body, 20 to 30 (usually 24 to 27) scales in the median row above the tail.

Colour dark brown, the sides and the terminal portion of the tail sometimes whitish, and all the scales in a few instances particoloured. Naked skin whitish.

Dimensions—A large male measured, head and body 21.5 inches, tail 20; basal length of a skull 4·1, greatest breadth 1·75.

Dyak name 'Tangiling.'

Hab. Mount Dulit 3000-ft. (C. Hose). Baram River (C. Hose).

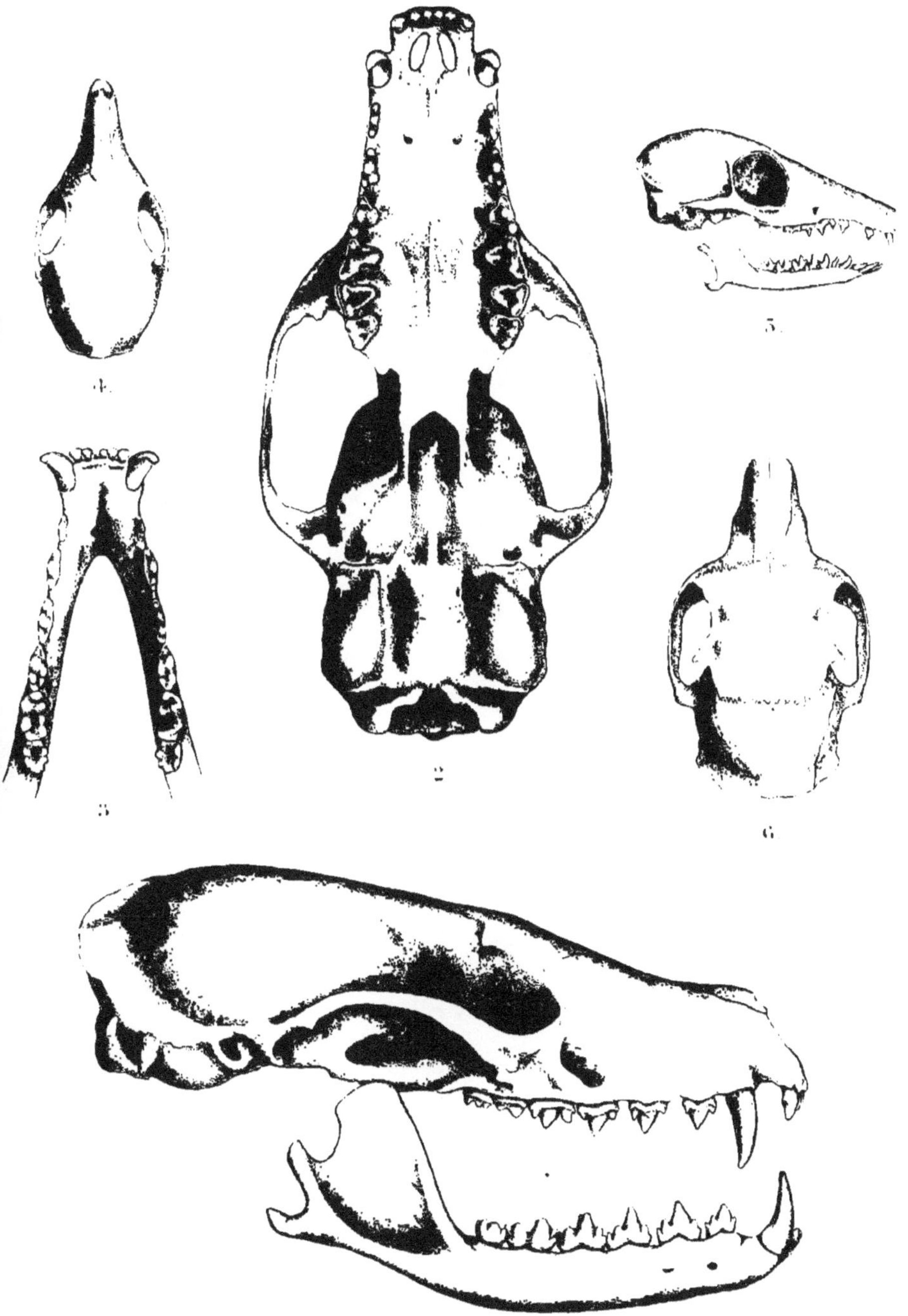

HINTS TO TRAVELLERS IN THE BORNEAN JUNGLE.

The work of exploration often proves a failure, owing, in a large measure, to the nature of the obstacles to be met with in the jungle; therefore, I have thought it advisable to mention some of the difficulties by which the traveller is always confronted, having had a fair experience of jungle life, and having, to a great extent, proved by constant practice the best methods by which such difficulties may be overcome. It is often the case that a little foreknowledge saves much inconvenience and discomfort, and so I have noted down a few matters that inevitably demand the attention of an explorer, in the hope that, by so doing, I may, to a slight extent, assist those engaged in this description of work, in making their journey more successful than it might otherwise be.

Do not encumber yourself with such baggage (necessary in some countries) as canvas tents, hammocks, mackintoshes or heavy boots. A hut—known to the Kayans by the name of *Sulap*, and to the Dyaks by the word *Lanko*—is easily made in the jungle. It is always advisable to have it raised about 2 feet off the ground, and the roof is made either of palm leaves or the bark of the "garungang" or "garu" tree, stripped from the trunk in large squares. So short a time is necessary for the erection of these structures, that, supposing the halt to be made at 4.30 p.m., everything can be in readiness for the night before darkness sets in. Caves, as resting places, are to be avoided, as they are cold and damp, and soon produce fever. In selecting a spot whereon to camp for the night, look immediately above and see that no dead wood is overhanging the hut.

Bathing in any of the rivers is attended with great risk, as all the Borneo rivers are infested with crocodiles, and natives are constantly being seized by these animals. Use a bucket and bale the water over yourself. On the mountains, about 4,000 feet up, the change in the temperature of the water is so great, that the bath, though very cool and refreshing, if too much indulged in during the day, soon produces ague.

It rains most days on the mountains. The best time to go out shooting is the very early morning, as soon as it is light, and before the mist rises. By nine o'clock the mist has risen from the rivers, and often nothing can be done for hours but wait for it to clear, an event which takes place about two o'clock in the afternoon.

Keep your matches in a tin box, or they will soon become useless from the damp ; you will then have to resort to a Dyak means of making a fire, which is obtained by manipulating a contrivance, consisting of two pieces of wood, called a "sukan." Never take matches that will strike *only* on the box, as the prepared surface on which the matches are ignited soon becomes damp, and the matches are useless. Wax matches are better adapted for the jungle than the wooden, as they are not so easily extinguished. If in a strong draught, fray a portion of the wax match before striking, as the larger flame which results is not easily put out.

If your gun caps are damp, or if your stock of them is exhausted, and you have no means of firing your muzzle-loader guns (good muzzle-loaders are by far the best for the natives), a match-head makes an excellent substitute. Cut off the head of a match, and fix it on top of the nipple of the gun ; pull the trigger in the ordinary way, and the gun is discharged as if by a cap.

Never allow the Kerosine oil—which is necessary for you to take for light-giving purposes—to be placed anywhere but in the bows or stern of your boat ; failing this precaution, you will soon detect a flavouring of this nauseous oil in your food. Servants are not at all particular in a cramped boat as to where they put the lamps, oil, etc., and as likely as not, you will see the bottle of oil in the box with the potatoes, or rubbing up against your small store of bread. A small lamp is often placed in a saucepan to be out of the way, but a trace of it will be found in everything subsequently cooked in that not-too-well-washed utensil, and only disappears after constant scrubbing. By having the lamps, oil bottles, etc. tied in the bows of the boat at the start, much unpleasantness can be avoided.

It is not necessary to carry water vessels in the jungle, for there are abundant ways of obtaining all the water you require, and there is hardly a place in Borneo where water is not procurable. The tough, skin-like, outer leaf of the nibong palm makes an excellent bucket, called "Upeh," or, failing that, the bark of a tree will always serve for a water vessel.

The cabbage of the nibong palm, known as "Umbut," is quite wholesome, and very useful as a vegetable.

When on the mountains, and unable to meet with water for immediate wants, you may always quench your thirst by cutting off lengths of many of the creepers, as these contain a quantity of water quite fit for drinking. The Dyaks speak of these creepers as "akar ai."

The food requirements of your men on an expedition are very simple, comprising rice, salt fish and salt, and tobacco to smoke. The proper allowance for each man per month is as follows:—Seven "*gantangs*" (35 lbs.) of rice; five "*katties*" (6 lbs.) of salt fish; half a "*katty*" (10 ozs.) of tobacco; and a small quantity of common salt. Whenever the opportunity offers, the fish should be put out to dry, or it will soon get soft and unpleasant. The rice should always be contained in long narrow bags, as in this form a man can carry a larger quantity. Cut an ordinary big rice bag down the middle lengthways, and sew up the side of each half, thus making two long bags, each holding about 50 lbs. of rice, which is as much as a man can carry. In buying rice for your men, it is a good rule to remember that eight men consume a big bag of rice in sixteen days.

Mosquitoes are always troublesome, but above 2,000 feet they usually disappear. Sandflies are at their worst on bright moonlight nights, and during the day on the rapids. Mosquito curtains, of sufficiently close texture to keep out the sandflies, are always necessary on levels below 4,000 feet; without such nets, you will be so worried by these pests, that sleep is out of the question, and your blood will soon be in a state of fever. When passing through the jungle you may often come to a flat open space, measuring from ten to fifteen yards across, and usually covered with moss and dead leaves. This is where the cock birds of the argus pheasants do battle, and these spots are called "balai kruai" by the Dyaks. It looks a very inviting place to rest for a few minutes, but it is really one to be shunned, for the ground is covered with small ticks from the birds, which cause the traveller much annoyance, though he may not be made aware of their presence till he camps for night.

The nests of the wild pigs, made of small boughs and twigs bitten off by the sows, often seen in the jungle, are also infested with noxious insects.

Land leeches of two varieties abound in the forest. The variety with a yellow stripe down the side inflicts the most painful bite, and a sore once caused by it is not healed without much trouble. If you sit down in the jungle for a few minutes and observe, you will see the leaves in motion all about your neighbourhood, and soon will discover that numberless leeches are making their way towards you. I have sometimes on halting taken off as many as twenty from each foot when I have not troubled to remove them whilst on the march. Never pull the leeches off, if you can help it, as they take with them a piece of skin; but squeeze some tobacco juice on them, and they will immediately fall away.

Bridges are easily constructed, either by felling trees or by tying bamboos across the river. If a river is so broad and deep as to be unfordable, get one of the party to swim across with an axe 'biliong,' and then fell a big tree on either side, selecting those about 130 feet in height, and taking care to make them fall up stream. The branches of the two trees will be locked together by the force of the current, and you can then walk down the trunk of one, cut your passage through the branches, and up the trunk of the other. Always take a 'biliong' with you, as it is a most useful tool.

Make-shift boats, called by the Dyaks 'utap,' may be very quickly constructed of the bark of a tree.

ALPHABETICAL INDEX

OF SCIENTIFIC NAMES.

M (continued).

PRINTED AND PUBLISHED BY

EDWARD ABBOTT, "EXPRESS" OFFICE, DISS, NORFOLK.

www.ingramcontent.com/pod-product-compliance
Lightning Source LLC
Chambersburg PA
CBHW020313090426
42735CB00009B/1324

* 9 7 8 3 3 3 7 2 4 5 6 8 9 *